REVISED EDITION

All About COTTON
A FABRIC DICTIONARY & SWATCHBOOK

WRITTEN & ILLUSTRATED BY JULIE PARKER

Revised edition
Copyright© 1998 by Julie Parker

Book design, cover design, text and all illustrations by Julie Parker.

Fabric Reference Series, Volume II

All rights reserved.

No part of this book may be reproduced in any form without permission in writing from the publisher, except by a reviewer, who may quote brief passages in review.

All inquiries should be addressed to:
Rain City Publishing
P.O. Box 15378
Seattle, WA 98115-0378
Phone: 206-527-8778
Fax: 206-526-2871
E-mail: RainCityPu@aol.com

Printed in the United States of America.

ISBN 0-9637612-3-4
LCCN: not yet known

Contents

- Introduction 6
- Fiber properties 9
- History & industry 12
- Shopping for cotton 18
- Caring for cotton 20
- Fabric definitions 23
- Mail-order sources 108
- Glossary of cotton terms 111
- Glossary of finishes 115
- Index 118

40 COTTON FABRICS

- batik 24
- batiste 26
- broadcloth 28
- calico 30
- canvas 32
- chambray 34
- chenille 36
- chino 38
- chintz 40
- corduroy 42
- damask 44
- denim 46
- dotted Swiss 48
- double knit 50
- drill 52
- duck 54
- eyelet 56
- flannel 58
- gauze 60
- gingham 62
- interlock 64
- jersey 66
- lawn 68
- Madras 70
- monk's cloth 72
- muslin 74
- organdy 76
- osnaburg 78
- oxford cloth 80
- piqué 82
- plissé 84
- poplin 86
- sateen 88
- seersucker 90
- shirting 92
- terry cloth 94
- ticking 96
- twill 98
- velveteen 100
- voile 102

TYPES OF COTTON

Combed cotton 27
Cotton/polyester blends 65
Egyptian cotton 69
Indian cotton 71
Naturally colored cotton 79
Peruvian cotton 29
Pima cotton 28
Sea Island cotton 93
Upland cotton 31

TYPES OF YARN

Chenille yarns 37
Dyed yarns 62
Stuffer yarns 82
Voile yarns 102
Yarn sizes 103

MISCELLANEOUS NOTES

Fiber identification 19
Industry trends 17
Manufactured fibers 8
Natural fibers 8
Stain removal 22
Thread count 75
When to dry clean cotton 101

SPECIAL FINISHES

Brushed cotton 58
Caustic-soda finishes 85
Durable-press finishes 92
Embossed cotton 84
Flame-retardant finishes 59
Flocked fabrics 48
Glazed finishes 40
Mercerized cotton 26
Preventing mildew 33
Shrinkage busters 39
Sizing 74
Stain-repellent finishes 41
Starched fabrics 80
Wrinkle busters 91

PRINTING & DYEING

Cotton prints 30
Crisp finishes 77
Greige goods 78
Indigo blues 46
Liberty prints 68
Piece-dyed fabrics 63
Resist dyeing 24
Setting the dye 25
Vegetable dyes 70

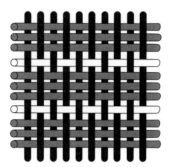

WEAVES & KNITS

Basket weave 73
Broken twills 99
Corduroy's wales 43
End-and-end 34
Herringbone twill 98
Jacquard weave 45
Knit fabrics 51
Lace construction 56
Lappet weave 57
Leno weave 61
Looped pile 95
Piqué weave 83
Plain weave 35
Rib knits 50
Rib weave 87
Satin weaves 89
Semi-basket weave 81
Tube knits 67
Twill weave 53
Unstable weaves 72
Warp-faced twill 52

TO MY PARENTS
To my parents,
Bruce and Shirley Parker,
who believed in my work
and encouraged me
to reach for the stars.

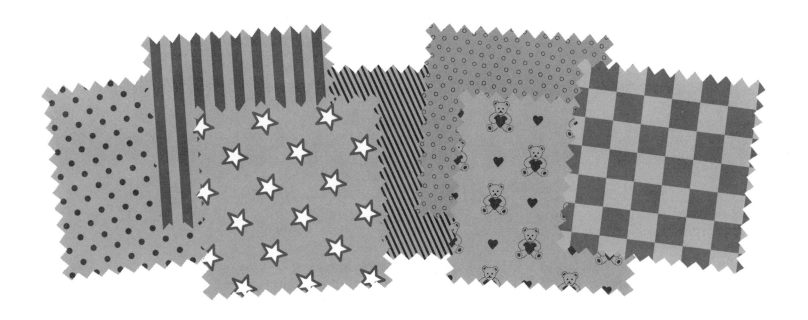

THE WORLD OF COTTON

FIBER PROPERTIES
HISTORY & INDUSTRY
SHOPPING FOR COTTON
CARE OF COTTON

KING COTTON

Cotton is truly the king of textiles — used the world over to make comfortable, affordable clothing that looks good and is easy to clean.

The popular natural fiber is used to make all kinds of fabrics, from fuzzy flannels to sturdy denims, in order to make all kinds of clothing, from soft pajamas to rugged overalls.

If you look around the house, you'll find cotton in almost every room, as bath towels and rugs, bed sheets and blankets, table cloths, napkins, dish towels, upholstery, window trimmings and throw pillows.

In the United States, cotton outsells all other fibers combined, with nearly 60 percent of the retail market, based on total fabric weight. Its nearest competitor is polyester, and much of that is sold as cotton/polyester blends.

Cotton hasn't always been so dominant. In the early 1800s, wool prevailed worldwide, with nearly 80 percent of the market. Cotton was third, behind linen. By 1900, cotton had grabbed wool's commanding share of the market, and it has been a title contender ever since.

In the 1970s, cotton was outpaced by synthetic fibers and the allure of wash-and-wear fabrics, which were introduced in '50s and came on strong in the '60s. Cotton staged a comeback in the 1980s, fueled in part by a renewed interest in natural fibers and by the increased casualness of corporate America.

Consumers of the '90s have demonstrated a willingness to pay extra for natural fibers, especially when it comes to cotton. And while some purchases may be influenced by catchy advertising jingos and trendy fashions, there are good reasons for sticking with cotton.

Cotton cleans up

When it comes to washability, cotton has no competition. It is the whitest, cleanest natural fiber and is easy to launder. It withstands high temperatures, even boiling, so

COTTON'S ASSETS

Cotton is a comfortable, easy-care fiber. Here are some qualities that make it desirable:

▲ **Washability**
Cotton is unmatched in washability. It is easy to launder in hot or cold water and can be tossed in a hot dryer or hung to dry without any damage to the fabric.

▲ **Detergents and bleaches**
Cotton is not harmed by strong detergents and may be bleached safely with ordinary household bleaches such as Clorox and Purex.

▲ **Strength and durability**
The moderately strong fiber gets stronger when it gets wet. It is not as strong as linen, but is tougher and more durable than rayon.

▲ **No bugs**
Cotton resists moths and pesky bugs that eat other fibers.

▲ **Absorbency**
Cotton can absorb up to 15 percent of its weight in moisture. It absorbs and releases perspiration quickly, it dyes easily and it can be bleached to a clear, bright white.

▲ **Coolness**
Cotton is porous, which allows the skin to breathe, making it especially comfortable in the summer. Crisp, clean fabrics look and feel cool.

cotton fabrics can be washed and sanitized in very hot water.

Cotton is highly resistant to alkalis, so it can be washed with almost any detergent or laundry aid. Cotton also survives bleaching very well, and most fabrics can be ironed at the highest temperature setting because cotton does not scorch easily.

Cotton gains strength when it gets wet, so it withstands abuse from washing machines, including those industrial behemoths used for commercial purposes. It can be tossed into a hot dryer or hung to dry without harm.

This all adds up to a fiber that is welcome wherever cleanliness is desirable — hospitals, restaurants, hotels, you name it, if you want it to be clean, over and over again, you want it to be cotton.

If you want to stay cool, cotton's your fiber again. It's comfortable to wear because it breathes and won't stick to your skin. On hot days, there is no substitute for a cool shirt of smooth, crisp cotton. That's why cotton is the universal choice of people who live in warm climates.

Cotton makes terrific underwear because of the fiber's high comfort level. Cotton absorbs moisture from the body and transmits it through the fabric so it can evaporate.

Cotton doesn't collect static, so it won't cling like synthetics do. And it doesn't rub people the wrong way, like wool sometimes does, so it can be worn comfortably next to the skin.

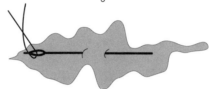

Working with cotton

Cotton is versatile and abundant. Fabrics vary in weight, quality and construction. Delicate lawn, sheer voile, plush corduroy, crisp organdy and industrial-strength duck are all made from cotton.

Cotton is available at most fabric stores, even when other natural fibers are scarce. As an added bonus, the resurgence of quilting has produced an endless bounty of cotton prints and solids in every color imaginable, sold at shops that sell nothing else.

Cotton can be blended with other

COTTON'S LIMITATIONS

Cotton is probably nature's most practical fiber. But the popular fiber has its limitations:

▼ **Drapeability**
Cotton does not have the body or suppleness for good drapeability, but this quality can be improved with special finishes.

▼ **Shrinkage**
Loosely woven cotton fabrics and knitted goods tend to shrink a lot, especially the first they are washed. Preshrinking is recommended for all cotton fabrics.

▼ **Wrinkles**
Cotton is not very resilient, so it wrinkles easily. The fiber has little natural elasticity and won't stretch much without breaking.

▼ **Sunlight**
Cotton fiber turns yellow and loses strength when exposed to sunlight for an extended period of time.

▼ **Mildew**
Cotton mildews rapidly when left in a damp condition. Sized fabrics are especially vulnerable. Heat speeds the rate of deterioration.

▼ **Perspiration**
Acid perspiration causes cotton to deteriorate slightly and affects the color, causing staining. Cotton should be cleaned frequently.

fibers, such as silk, linen or rayon, to change the look and feel. It is most often blended with polyester, usually to cut costs and reduce wrinkles.

Cotton is ideal for beginning sewers because it's easy to work with and doesn't cost very much, so it won't hurt too much if you need to buy another yard or two along the way. Even the most experienced sewers can be intimidated by the thought of ripping into a costly silk or wool.

And while beginners can cut their teeth on cotton, it provides enough challenge and variety for seasoned veterans as well. A delicate batiste inspires the detailed precision of tiny hand-stitched pintucks, while a wedding gown of dotted Swiss can be playful or sophisticated, and the subtle luster of sateen adds spark to the swirl of a long, full skirt.

Whether you're looking at cotton for the first time or the hundredth, you'll have to agree that this puff of fluff from the seed pod of a plant is a pretty useful fiber. And the more you get to know cotton, the easier it is to understand why — in this age of high-tech manufactured fibers — it remains king of textiles, the most widely used fiber in all the world.

NATURAL FIBERS

Natural fibers exist naturally in fiber form. There are two sources: plants and animals. Natural fibers are comfortable to wear because they are porous and absorbent, so they allow the skin to breathe and shed perspiration.

Plant fibers, also called vegetable fibers or cellulose fibers, are made up mostly of cellulose, a carbohydrate. Cotton and kapok are examples of **seed fibers**, which grow as hair or fuzz from seeds. Most seed fibers are short and soft, and some are too short to spin into yarn. Linen, hemp and ramie are **bast fibers**, which come from the stem or stalk of a plant. Bast fibers are longer, stiffer and more lustrous than seed fibers. Sisal and abaca are examples of **leaf fibers**, which are long, strong and stiff.

Animal fibers are called protein fibers because they consist mostly of protein. Wool comes from sheep and specialty wools come from camels, goats and similar animals. Wool fibers vary in length, but they tend to be short and must be twisted and spun to make yarn. Silk is the only natural filament fiber, unreeled from the cocoon of a moth, whose larva is called a silkworm.

MANUFACTURED FIBERS

Silk cocoons

Manufactured fibers do not exist in nature — they were cooked up in a chemistry lab. Usually, a manufactured fiber begins as some form of liquid that is squeezed through a tiny hole and allowed to harden, forming a long filament. The fiber's shape can be changed to produce different characteristics, such as crimp, elasticity or luster. Fibers also can be cut into shorter lengths and spun into yarn.

Cellulosic fibers are made from regenerated plant matter, usually cotton linters or wood pulp, that is reduced to liquid form. Cellulosic fibers breathe like natural fibers, which makes them comfortable to wear, but they are not as strong as synthetic fibers. Cellulosic fibers include rayon, acetate, triacetate and lyocell, currently sold as Tencel®.

Synthetic fibers are made from petroleum byproducts, natural gas, alcohol and coal. They don't absorb moisture, so they are less comfortable to wear right next to the skin, but they are strong and durable. Synthetic fibers include nylon, polyester, acrylic, modacrylic, spandex and olefin.

UNDER THE MICROSCOPE

Cotton is a fluffy, soft fiber that grows from the surface of seeds in the pods, or bolls, of a bushy plant or small tree. Different varieties of cotton grow in warm climates all over the world. They have somewhat different characteristics, but they all belong to the genus *Gossypium* of the *Malvaceae* family.

The main ingredient of all plant fibers is cellulose, a carbohydrate. Raw cotton is about 90 percent cellulose and 6 percent moisture. The rest is natural impurities. After processing, cotton is 99 percent cellulose.

Raw cotton is creamy white. As it ages, it becomes more beige. If it rains just before harvest, the fiber turns a bit gray. White is preferred. In addition to whiteness, quality is judged by the fiber's strength, length, fineness and maturity. Long fibers are finer, stronger and more desirable because they can be made into smoother, softer, stronger, more lustrous fabrics.

Depending on the variety, cotton fibers vary in length from just under an inch to about $2\frac{1}{2}$ inches. Most cotton fibers grow about 1 to $1\frac{1}{2}$ inches long. They are divided into five groups: extra long staple, long staple, medium, short and extra short.

(**Staple** is the textile industry's term for fiber, used to describe shorter lengths that must be spun and twisted to make yarn; **filament** is used to describe long, continuous strands of silk or manufactured fiber.)

The diameter of cotton fibers is measured in microns. One micron is equal to $1/1,000$th of a millimeter. Long fibers are finer — about 15 microns — while short fibers are coarser, up to 20 microns. Most cotton fibers are about 18 microns.

Variations among fibers occur because of differences in weather, climate, soil, fertilizers, harvesting techniques and insect damage, such as the boll weevil.

From seed to fluff

The fiber grows from the seed as a hollow tube that is about 1,000 times as long as it is thick. The fiber is made up of a **cuticle, primary wall** and **secondary wall**, all surrounding the **lumen**.

The cuticle is a wax-like film covering the outer primary wall. The primary wall surrounds the secondary wall, which is made up of layers of cellulose that form growth rings on the fiber.

Inside the secondary wall is the lumen, a central canal that carries nutrients during growth. After the fiber has matured, dried nutrients in the lumen appear as dark spots when viewed with a microscope.

Cotton fiber is relatively smooth and straight, but a natural twist, or **convolution**, develops as the fiber matures. The ribbon-like twist enables the fibers to stick to one another, so that in spite of its short length, cotton is one of the most spinnable fibers. Long fibers

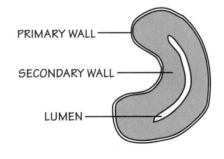

CROSS-SECTION OF MATURE COTTON FIBER

have about 300 convolutions per inch; short fibers have 200 or less.

The layers of the secondary wall are composed of **fibrils** – bundles of cellulose chains – arranged spirally. The spirals reverse direction at regular intervals. The reverse spirals play a key role in the fiber's twist, elasticity and stretch. They also are weak points, being 15 to 30 percent weaker than the rest of the fiber.

REVERSE SPIRALS OF COTTON FIBER

The fiber changes shape as it grows. Immature fibers are rounder, with a large central canal to carry nourishment. When the boll opens, the fiber dries out and the central canal collapses. Mature fibers are U-shaped.

Each boll contains mature and immature fibers. The proportion of one to the other causes problems in processing, especially in spinning and dyeing. Machine-picked cotton contains more immature fibers than hand-picked cotton because the cotton is picked all at once instead of waiting for each boll to ripen.

Cotton's personality

Cotton absorbs moisture, which makes it extremely comfortable to wear. Absorbent fibers "breathe" by pulling moisture away from the body and releasing it into the air through evaporation.

Cotton can hold up to 15 percent of its weight in moisture before it feels wet. That's why cotton stays cool on a hot summer day.

Synthetic fibers also can remove moisture from the body, but it sits on the surface of the fabric instead of passing through. On a hot day, a synthetic fiber will plaster itself to your skin because perspiration is trapped on the inside the garment and can't evaporate.

Absorbent fibers are the easiest to dye, especially with dyes that are dispersed with liquids. Cotton dyes easily and fabrics come in all colors of the rainbow.

Absorbent fibers are static-free, so cotton garments won't spark, crackle or cling, unless they are dried too long in the clothes dryer, a problem that is easily fixed by hanging the garment in a steamy bathroom, where it will drink up the lost moisture.

Strength and durability

Cotton has medium strength and is stronger when wet, which means it can stand up to rough handling in the wash. Short fibers are not as strong as the longer fibers used in combed yarns, so a combed cotton fabric will be stronger and more durable than a comparable fabric made of carded yarns. Likewise, a thin fabric is less durable than a heavier one – batiste and lawn will wear out faster than denim and corduroy.

Cotton is only moderately flexible. A flexible fiber increases a fabric's durability because it can be bent or folded repeatedly in the same place without breaking. Wool and silk are more flexible than cotton, while linen is less so.

Fabrics made from flexible fibers are more fluid than fabrics made from stiff fibers. Cotton fiber does not produce an especially beautiful drape, but some fabrics are treated with a special finish to enhance drapeability.

Stretch and recovery

All fibers stretch. They vary in how much they stretch before breaking, and in their ability to return to size. The length of time a fiber is kept in an elongated position also affects the rate of recovery.

Some fibers stretch easily and bounce right back, like the snap of a rubber band, while others resist stretching and creep back to size. Some fibers never fully recover.

Cotton fiber is not very elastic. It breaks after moderate stretching and does not recover well, so fabrics tend to stay stretched in areas of stress, such as elbows and knees.

Cotton has very low resiliency, so fabrics that are folded or crushed will stay that way. Creases can be pressed in easily, and wrinkles can be pressed out, but garments tend to develop wrinkles while being worn.

Shrinkage

Cotton has a reputation for shrinkage that is badly misplaced. Shrinkage is not a characteristic of cotton fiber, but the result of fabric that is mass-produced in our high-tech, high-speed world. If you work enough with cotton, you will discover that shrinkage is not consistent from one fabric to the next, and in many cases, it simply does not occur.

That's because cotton fibers don't really shrink. If you wash a strand of cotton fiber, it will always return to the same size when dry.

Most shrinkage is caused by the way we spin, weave and finish fabric. At the mill, fibers get stretched when they are spun into yarn and again when they are placed under tension on a loom. When the tension is removed, the fiber will relax a little or a lot, depending on how much it was stretched and how long it was forced to stay that way.

Most cotton fabrics are kept under tension from start to finish, so distorted fibers don't relax until the fabric is washed the first time, usually at home by the consumer. Once this relaxation shrinkage is removed, the fiber may continue to stretch and relax a bit each time a garment is worn and washed, but the size should remain stable.

Cotton knits, loose weaves and twills such as denim and chino are especially vulnerable to shrinkage. Fabrics washed and dried using high temperatures will shrink the most.

Clothing manufacturers minimize the effects of relaxation shrinkage by prewashing garments or applying special finishes. Home sewers can save themselves a lot of shrinkage headaches by preshrinking every cotton fabric before cutting into it.

Chemical reactions

Cotton stands up well to alkalis, such as laundry detergent, but is vulnerable to acids. Its reaction to both alkalis and acids is increased with higher temperatures.

Strong acids will destroy cotton fibers. Mild organic acids, such as lemon juice and household vinegar, can cause harm. Rinse spills immediately with cold water.

The acid in human perspiration weakens cotton fibers and may cause discoloration. To avoid this, cotton should be cleaned frequently.

Chlorine bleach is safe to use as long as it is used properly. Too much bleach weakens cotton fibers. Don't use bleach every time you wash — save it for the removal of stubborn spots and heavier-than-usual grime.

Organic solvents won't harm the fiber, so it can be dry cleaned safely.

Cotton's weaknesses

Window coverings and outdoor fabrics deteriorate quickly because cotton loses strength when it sits in the sun too long. Sunlight causes white and pastel cottons to yellow.

Damp cotton mildews rapidly. Starched and sized fabrics are especially vulnerable, and heat speeds the rate of deterioration.

Moths and carpet beetles prefer wool and silk, but silverfish love cotton, especially if it has been starched or sized.

Cotton ignites easily and burns rapidly. It will continue to burn after the source of ignition is removed. Anyone wearing thin cotton clothing should be careful around hot stoves and fires.

COTTON HISTORY & INDUSTRY

The exact origin of the cotton plant is not known, but the fluffy white fiber has been cultivated and used to make cloth for at least 5,000 years. Evidence suggests that cotton grew in India and Mexico more than 7,000 years ago. It may have existed in Egypt as early as 12,000 B.C.

Fragments of cotton fabrics have been found by archeologists in Mexico (from 3500 B.C.), in India (3000 B.C.), in Peru (2500 B.C.) and in the southwestern United States (500 B.C.).

The first documented evidence of cotton is found in the writings of Greek historian Herodotus. In 445 B.C., after a trip to India, he wrote about trees that grew woolly fleece superior to that of sheep, from which the Indian people made cloth.

The Greeks didn't begin to grow cotton until the second century A.D. and it was much later before cotton spread to other countries.

Spain was introduced to cotton at the turn of the 10th century. By the 13th century, a thriving cotton industry existed in Barcelona for the purpose of making sails.

In 1492, Christopher Columbus found cotton growing in the Bahamas. A few years later, Spanish settlers planted cotton in nearby Florida. Other Spanish explorers discovered finely woven cottons in Peru, Mexico and what is now the southwestern United States.

England got its first taste of cotton in the 13th century, but the use of cotton there did not become widespread until the 16th century.

In 1607, British colonists in Virginia planted seed from the West Indies, the first commercial planting of cotton in North America. By the end of the century, England's wool industry found itself competing with cotton exports from America.

England responded by passing laws forbidding the use of cotton, including one that promised a stiff fine for anyone caught wearing the stuff. The prohibitions didn't last — by 1730, the first cotton yarn had been spun by machine in England.

The industrial revolution

By 1791, U.S. cotton growers were shipping 400 bales a year to Europe, most of it to England. By 1810, the number had increased to 180,000 bales. During the same 20 years, a series of spinning and weaving inventions mechanized fabric production in England. For the next 50 years, England's cotton textile industry continued to grow rapidly.

Back in the United States, the Industrial Revolution launched a number of mechanical innovations that improved cotton production. The most significant was the cotton gin in 1793, a hand-operated gadget that separated cotton fiber from seeds, led to the dominance of the South as a cotton producer and turned the fledgling United States into a world factor.

Eli Whitney lost money on his invention, but the gin eventually was accepted by cotton growers. It was directly responsible for the sudden growth in cotton production, which fed a massive trade business between North America and Europe and led to the industrialization of both continents.

Entrepreneurs began to build factories to spin and weave cotton, most of them in the north where water power was abundant. By 1810, there were 226 mills in New England, which soon rivaled England in the production of cotton fabrics.

As manufacturing techniques improved, the demand for cotton exploded in Europe and other parts of the world. The United States struggled to keep up. The biggest problem facing cotton growers was the rapidly increasing number of workers needed to harvest the crops, which were picked by hand.

The use of slaves had been declining in the United States, but the financial allure of the cotton market soon reversed that trend.

Slaves and civil war

Before Eli Whitney's invention, the principle of slavery had fallen into disfavor. States had passed laws against slavery and, in 1787, the entire South had voted unanimously against the importation of slaves.

The cotton gin turned the South on its ear and the acceptance of slave labor staged a comeback. By 1810, there were more than 1 million slaves in the South.

By 1860, cotton was America's leading export and "King Cotton" was a common expression all over the country. Annual production had reached 4.5 million bales, two-thirds of the world's supply. Meanwhile, the slave population had expanded to more than 4 million, leading to four years of bitterly contested civil war.

After the Civil War ended in 1865, the South struggled to rebuild its cotton industry without slaves. New machinery inspired entrepreneurs to build cotton mills throughout the South, where most of the U.S. industry remains today.

Until 1977, the United States was the world's biggest producer of cotton. Today, it competes with China for the top spot. Cotton is cultivated across the South, from Virginia to California.

Other important cotton growers include Egypt, South America (especially Peru and Brazil), Central America, Mexico, India, Africa and the West Indies. The countries that make up the former Soviet Union also grow a lot of cotton.

Types of cotton

Usually, cotton is identified by the plant type and by the name of the country or region where it grows. The most common type of cotton is Upland, which is further identified by its source, such as India or Brazil.

The source is important because Upland cotton varies in quality and characteristics, depending on the growing conditions. For example, Upland cotton from California's San Joaquin Valley has fibers that are longer and stronger than Upland varieties grown elsewhere.

Most of the U.S. crop is Upland cotton, which is used to make a wide variety of fabrics. Most of the rest of the U.S. crop is Pima, a superior cotton cultivated in Texas, Arizona, New Mexico and California. Pima has very fine, strong, lustrous, extra long fibers and is used to make yarns and fabrics of excellent quality.

Egyptian cotton is regarded as the finest cotton in the world, with extra long, fine fibers. It is used to make very fine yarns and expensive, high quality fabrics. Bed sheets of Egyptian cotton usually have a high thread count and may cost as much as $500 per set.

Peru also is known for its fine cotton, most of which is cultivated on small farms and picked by hand.

Sea Island cotton from the West Indies can hold its own against Egyptian and Peruvian fibers, but it is not very common because it is difficult to cultivate.

From seed to fiber

Cotton thrives in warm, humid climates and sandy soil. Under the most favorable conditions, it needs six to seven months of continuous warm weather with 3 to 5 inches of rain during growth and no danger of frost, which damages the plants. Cotton also is grown successfully in the irrigated fields of dry climates, such as the southwestern part of the United States.

Most of the U.S. crop is planted in March and April. In June or July, the plants burst into bloom with creamy white or pale yellow flowers, which turn to pink, lavender or red by the next day. Within three days, the petals fall off, leaving the immature cotton boll, which is really a seed pod. As the fibers grow from the seeds, the pod expands until it is about an inch in diameter and $1\,1/2$ inches long.

During this time, the plant is vulnerable to attack by a number of insects, especially the boll weevil, which punctures the boll and lays its eggs inside. The larva eats the boll from the inside as it develops.

The cotton bolls grow to full size by August or September. Plants may be 3 to 6 feet in height. The bolls are ready to pick when they begin to burst with fleecy white fiber.

Harvesting

Not all cotton bolls open at the same time. Only those that burst open to expose the fiber are ready for picking. A hand-picked field must be harvested several times. Fields picked by machine are not harvested until all the bolls have matured. The plants are sprayed with a chemical, causing the leaves to wither and fall off. The bolls are then stripped from the plants by a mechanical picker.

A hand picker can gather about 15 pounds of cotton in one hour. A mechanical picker can harvest as much as 650 pounds an hour, but the machine picks up a lot of waste with the fleece and the cotton is not as clean as hand-picked cotton.

Hand-picked cotton is better in quality and more uniform, but most fields are harvested by machine because workers are scarce and the cost of labor is high.

Ginning and baling

After the cotton is harvested, it must be cleaned to prepare it for spinning. Raw cotton contains dirt,

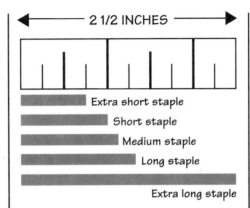

pieces of leaf and other undesirable material. The seeds alone make up about two-thirds of the weight.

The cotton gin removes most of this extraneous matter with rows of revolving teeth that pull the cotton lint away from the seeds and other material. The ginned cotton is then graded and shipped to the mill.

Staple length, or fiber length, is one of the main factors used to determine the cotton's grade. The fiber length is affected by the type of cotton and growing conditions. There are five main divisions:

Extra short staple describes any fiber shorter than 3/4 inch. It is not very suitable for spinning and is used to make batting and stuffing.

Short staple cotton is 3/4 to 1 inch long and is used to make coarse, less expensive fabrics. Most Indian, Chinese and other Asiatic cottons fall into these two groups.

Medium staple cotton is 1 inch to 1 1/8 inches long. Many American Upland cottons fall into this range, along with Upland varieties from India and Pakistan. **Long staple** cotton is 1 1/8 to 1 3/8 inches long. Better varieties of American Upland cotton fall into this range.

Extra long staple includes the best cottons with fine, strong fibers that range from 1 3/8 inches up to about 2 1/2 inches. Egyptian, Pima and Peruvian fall into this group.

From raw cotton into cloth

At the mill, the ginned cotton is processed through a series of machines that pick out remaining seeds and dirt and straighten the fibers to prepare them for spinning into yarn. This is where the cotton is **carded** and, when a higher quality is desired, **combed**.

The carding machine has revolving cylinders spiked with wire teeth that pull out dirt and straighten the fibers so the cotton can be spun into yarn. Combing repeats this process on a finer scale, removing short fibers and any remaining dirt. Most cotton is too short to be combed, so it is carded only.

After carding or combing, the bulk fiber is **drawn** into an increasingly thinner strand and **spun** into yarn by twisting it to hold the fibers in place.

The **weaving** of cotton fabrics is basically the same as any other fiber — the warp yarns are set up on a loom and interlaced with crosswise filling yarns. Because cotton is not especially strong, the warp yarns often are coated with starch or sizing to keep them from breaking under tension on the loom.

Knitting is the other method of turning yarn into cloth or directly into a finished garment. Fabrics or garments may be produced by a hand knitter using two needles or by a high-speed circular or flatbed knitting machine.

Since the 1970s, cotton knits have maintained a steadily growing share of the market, especially for leisure wear and active sportswear, because they are comfortable and inexpensive. Cotton T-shirts are a staple of the knitwear industry.

Final steps

When cotton fabric comes off the loom, it contains small bits of dirt and other foreign matter picked up along the way. It must be removed before the fabric can be printed, dyed or finished.

First, fabrics are brushed lightly to raise the ends of any protruding fibers. These are **singed** to remove them, either with heated plates or open flames.

The singed fabric is dipped at once into a chemical **desizing** bath, which douses any troublesome sparks or afterglow. Desizing also removes the sizing that was applied to the warp yarns before weaving, as well as any particles that remain in the cloth.

To achieve bright, clear, even color, fabric must be pure white, so most cottons are **bleached** to remove all traces of natural color. Finally, the fabric is washed and rinsed to get rid of any chemical residue left behind by these processes.

Fabric endures a lot of stress and strain during processing, so it must be **tentered** to straighten the grain and adjust the width. The fabric is stretched taut and dried while held in place with clips or pins, called tenterhooks. (The small holes in the selvages of most fabrics are from tenterhooks.) A properly tentered fabric has yarns that intersect at 90-degree angles.

Dyeing and printing

Dyeing can be done at any stage — as fiber, yarn, fabric or garment. Most cottons are dyed either as yarn or fabric, called **yarn dyeing** and **piece dyeing**. Both methods are economical and give good results.

Most cottons are printed using one of two methods: roller or screen printing. **Roller printing** is similar to newspaper printing — the design is engraved on copper or steel plates that revolve as yards and yards of fabric pass by at high speed.

Screen printing is an advanced form of stenciling. The design is etched into a screen and a colored paste is passed through the screen onto the cloth. It can be done using a flatbed screen and a mechanical squeegee, or with a rotary screen that passes color to the fabric from the inside of a revolving, etched cylinder. Some screen prints are made by hand, using small flatbed screens and hand squeegees.

Finishing

Most fabrics are treated with one or more finishes to alter the fabric's appearance and/or behavior. Some finishes are applied before the fabric is dyed or printed, while others are applied afterwards, using either a chemical or mechanical process. There are four basic types:

Permanent finishes change the fiber's structure and will last the life of the fabric. Mercerized cotton is one example. It is done to make the fabric stronger, easier to dye and more lustrous.

Durable finishes, such as durable press, last as long as the garment, but they become less effective with each cleaning.

Semi-durable finishes eventually wash out or wear off. Usually, these finishes are applied to fabrics that do not need frequent cleaning, such as upholstery goods. The glazed finish of chintz is an example.

Temporary finishes disappear the first time the fabric is washed. Such finishes include **sizing**, a starch-like substance that stiffens the fabric, and **calendering**, a mechanical finish that smooths the fabric, much like ironing on a larger scale.

TRENDS IN COTTON FABRICS

Cotton has been around for centuries, but the face of cotton is forever changing. As we prepare to flip the calendar to a new century, be on the lookout for:

- ✓ *Fine career suitings for women with the look of wool, made of cotton blended with worsted wool.*
- ✓ *Lightweight summer suitings for women, made of cotton blended with mohair and calendered for a smooth finish.*
- ✓ *Fine women's shirtings with intricate dobby patterns and two-ply yarns – one ply of cotton, the other of silk.*
- ✓ *Double-faced women's shirtings with comfortable cotton on the inside and lustrous silk on the outside.*
- ✓ *All-cotton fabric treated with a special silicone finish to make it look and feel like linen, without the higher cost of linen.*
- ✓ *Cotton/linen blends for men's and women's slacks, treated to resist wrinkling.*
- ✓ *Soft, mercerized cotton knits with subtle luster and increased sophistication.*
- ✓ *Sparkle knits of cotton blended with nylon for a shimmery, energetic look.*
- ✓ *Textured knits with three-dimensional patterns and interesting surface designs.*
- ✓ *Lightweight denims, textured crinkle looks and new finishes, such as sueding.*

HOT COTTON

Other uses

The cotton plant is the source of many other products. Linters are short fibers that remain stuck to the seed after the cotton is ginned. The seeds are ginned a second time to remove the linters, which are used to make rayon and acetate fabrics, plastics, paper, shatterproof glass, high-gloss lacquers, liquid cement, film, twine, wicks, carpets, surgical gauze and stuffing for mattresses and furniture.

The seeds are separated into hulls and meat. Hulls are used for paper, plastics, explosives, fertilizer and cattle feed. The oil is removed from the meat and used to make cooking oil, shortening, margarine, lubricant, paint and soap. The meat is used for livestock feed, flour and fertilizer.

These days, cotton is even being recycled. Levi Strauss & Co. uses denim scraps to make the company's stationary, which has a faint blue tint that reminds you of, well, denim.

SHOPPING FOR COTTON

Most stores carry a wide range of cotton fabrics and cotton/polyester blends, but the quality varies considerably. Here are some tips for judging a fabric's quality before you buy:

✓ Count the number of yarns per inch in both directions of the weave. Better fabrics have more yarns per square inch.

✓ Hold the fabric up to the light. The tighter the weave, the less light will shine through. Look for uniform spacing between yarns. Beware of fabrics that appear to be thinner in some areas than others.

✓ Inspect the twist of an individual yarn, preferably a yarn pulled from the lengthwise direction of the weave. A tightly twisted yarn is stronger and the fabric will be more durable. The presence of short, wispy fibers may be a sign of low quality yarns.

✓ Inspect the yarns for variations in thickness. They should be uniform in size and thickness, although they may not be the same size in both directions, depending on the weave. (Keep in mind that slubbed yarns and irregular weaves are part of the appeal of some fabrics, especially handwovens.)

✓ Scratch a small section of the fabric with your thumbnail to see if the yarns are distorted easily. Such fabrics tend to pull apart at the seams of close-fitting garments.

✓ Check the fabric to see if the lengthwise and crosswise yarns intersect at right angles. Beware of fabrics with bowed or skewed grain.

✓ Inspect printed fabrics to make sure the print is straight, especially in the case of stripes, plaids, checks and other designs with straight lines. The lines of the print should match the grain or torn edge of the fabric.

✓ Inspect printed fabrics to make sure the colors are registered, with no overlapping colors, blurred prints or white outlines around the motifs.

✓ Rub the fabric between your hands. A heavily sized fabric will feel chalky. It may appear to be coated with a filmy or powdery substance. Such fabrics usually are inferior.

✓ Check the fabric for loose yarns to see if it unravels easily.

✓ On twill fabrics, examine the angle of the diagonal twill line. A steep line is a sign of a tight, compact weave and a better quality fabric.

✓ Scrunch a small piece of fabric in your hand and release it. If the fabric remains wrinkled, it probably will wrinkle easily.

✓ Loosely gather up to a yard of the fabric in your hand to determine how it will work with a gathered or shirred style. Unroll a yard or more

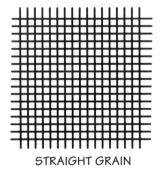 STRAIGHT GRAIN 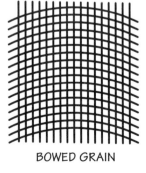 BOWED GRAIN 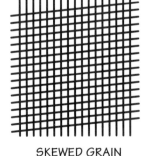 SKEWED GRAIN

of the fabric and let it hang freely to determine how it drapes and falls.

✓ Examine the fabric for one-way designs, luster, nap or other factors that may require special handling.

✓ Sniff the fabric for unpleasant odors that may not wash out.

✓ Compare the price with that of similar fabrics. Although price is one sign of quality or lack thereof, it is no substitute for a thorough look at the fabric. Some fabrics are true bargains and some are not.

The burn test

The burn test is a quick and easy way to determine the fiber content of unmarked fabrics.

Cut a small piece from the fabric and loosen a few yarns with your thumbnail. In a well-ventilated room, light a match and slowly move the unraveled corner of fabric toward the flame. Observe the reaction.

Move the entire sample directly into the flame and observe its burning characteristics.

Remove the sample from the flame and observe the reaction. Be prepared to extinguish any samples that continue to burn.

Let the sample cool and examine

THE BURN TEST

Cotton and linen
Does not shrink away from flame. Burns quickly with a bright yellow flame. Continues to burn when the flame is removed. Produces a small amount of fluffy gray ash. Smells like burning paper. To determine which is which, unravel a yarn. Cotton has short, soft, dull fibers. Linen has longer, stiffer, more lustrous fibers.

Silk
Curls away from the flame. Burns slowly with some melting. Burns very slowly when the flame is removed and may self-extinguish. Leaves a brittle, round, black bead that is pulverized easily. Smells like burning feathers.

Wool
Curls away from the flame. Burns slowly with some melting. Burns very slowly when the flame is removed and may self-extinguish. Leaves a lumpy, brittle, blistered ash that breaks easily. Smells like burning hair.

Rayon
Does not shrink away from the flame. Burns very rapidly with a bright yellow flame. Continues to burn when the flame is removed and leaves a creeping ember. Produces a small amount of ash or no ash. Smells like burning wood.

Acetate
Melts away from the flame. Burns with melting and continues to burn and melt when the flame is removed. Leaves a brittle, irregular black bead. Smells like vinegar.

Polyester and nylon
Fuses and shrinks away from the flame. Burns slowly and melts, usually with no flame. Stops melting when the flame is removed. Polyester leaves a hard, tough, round black bead and produces a chemical odor. Nylon leaves a hard, tough, round gray bead and smells like boiled string beans.

Acrylic
Shrinks away from the approaching flame. Burns rapidly and melts. Continues to burn and melt when the flame is removed. Leaves a hard, brittle, irregular black bead. Produces an acrid odor.

the characteristics of the ash or residue.

By this time, the burning sample will have produced a distinct odor. What does it smell like?

The chart on page 19 provides information on most of the major textile fibers and how they respond to this simple test.

Before you begin, you may want to practice by burning small samples of known fibers. Compare the results to each other and to the chart.

Keep in mind that the results of a burn test can be inconclusive, especially if more than one type of fiber is present, or if the fiber has been altered by a special finish. Pure, unadulterated fibers are much easier to identify than a mixed bag.

Cotton treated with a chemical or resin finish will burn differently than plain cotton, and may produce a chemical smell instead of the smell of burning paper. Such finishes often show up as luster on one side of the fabric, and the fabric may seem a bit stiff.

The type of weave also affects the results. A dense weave will be slower to ignite than a wispy gauze, which may literally go up in flames.

CARING FOR COTTON

Cotton soils easily, so it must be cleaned frequently to keep dirt and stains from establishing a permanent hold on the fabric.

It can be washed in hot, warm or cold water, but detergents lose some of their punch in cooler water, so warmer water is more effective, especially for heavily soiled cotton. Cold water, on the other hand, will minimize shrinkage and wrinkles.

Wet cotton is stronger than dry cotton, so most fabrics can take a beating from the normal agitation cycle of a typical washing machine, although a gentle cycle is preferred for knits and delicate fabrics.

The home laundry is recommended because cotton is not harmed by strong detergents or laundry aids, and the fiber can be sanitized, if necessary, in very hot water.

Professionally cleaned garments are preferred by some because cotton can be difficult to iron. There's nothing quite like the crisp, smooth appearance of a professionally laundered shirt, but repeated dry cleanings will turn white cottons gray.

Dry cleaning does not remove water soluble stains. Instead, the heat from cleaning, steaming and pressing may set a stain permanently. Many stains can be removed by a dry cleaner if they are pointed out by the consumer, but don't assume they will disappear on their own.

The yellow stains of perspiration, in particular, will be made permanent by dry cleaning, but they can be controlled at home with consistent use of a pre-wash product or bleach.

Using bleach

A little bleach, however, goes a long way. White cotton is routinely bleached to keep it bright, but too much bleach gradually weakens the fiber. Overbleached cotton tears easily when it gets wet, so it is prone to wear-and-tear in the wash.

Strong chlorine bleach, such as liquid Clorox and Purex, will kill mildew, bleach out stains and sanitize all-cotton fabrics, but it will fade the color of dyed fabrics and turn synthetic fibers yellow. Cotton/polyester blends should be treated with a weaker oxygen bleach, usually sold as a dry powder, such as Snowy or Clorox II.

Avoid chlorine bleaches on fabrics that have been treated with a resin finish to cut shrinkage or wrinkles. The fabric may turn yellow and the finish may be destroyed.

A good rule of thumb is to resist the use of bleach unless nothing else works, and be sure to rinse it out thoroughly.

Avoiding shrinkage

Many cotton fabrics and garments are likely to shrink the first time they are washed. The law of cotton shrinkage can be summed up with three simple statements:

1. Cotton knits shrink the most.
2. Heat exaggerates the results.
3. It is easier to allow for shrinkage than to try and prevent it.

Always preshrink cotton fabrics before cutting into them. Some fabrics will shrink more than others, and some may not shrink at all, but this simple step will end shrinkage nightmares.

The same cannot be said for cotton garments. The consumer is forced to guess about possible shrinkage and buy jeans that are too long, T-shirts two sizes too big and sweaters to be washed by hand.

Some garments are pretreated for shrinkage. Prewashed cotton is a good choice, because the garment has been washed and shrunk before you ever try it on. Mechanical and resin finishes also cut shrinkage, but they change the nature of the fabric and tend to wear off with age and repeated washings.

Shrinkage can be minimized by using cold water and avoiding the heat of a clothes dryer. Most shrinkage occurs at the end of the drying cycle, so it helps to remove the garment from the dryer while it is still damp. A dry garment will continue to shrink in the dryer, so don't overdo it unless shrinkage is your goal.

Beware of cotton garments with care labels that suggest washing in cold water and drying flat — it's usually an effort to avoid inevitable shrinkage.

Avoiding wrinkles

The bad news: cotton wrinkles easily. The good news: wrinkles are easily pressed out with an iron. The trick is to avoid wrinkles in order to minimize ironing. There are several ways to accomplish this:

Buy fabrics that resist wrinkling, such as corduroy, seersucker, knits, denim and velveteen.

Don't overstuff the dryer. Clothes that do not tumble freely are more likely to wrinkle. Remove clothes as soon as the cycle is finished, or continue to tumble with no heat.

Don't leave cotton garments wadded up in the clothes hamper. Smooth out wrinkles and fold them or hang them up immediately.

If all else fails, you may have to get out the iron. Cotton can take the heat, usually the iron's highest temperature setting. You'll get better results with a little steam, which can be generated by spritzing the fabric with mist from a spray bottle, placing a damp press cloth over the fabric or pressing the fabric while it is slightly damp.

Cotton/polyester blends require a lower temperature setting — a hot iron will scorch or melt the fabric. A light touch is all you need, because these fabrics don't wrinkle as much as all-cotton types.

Storage of cotton

Cotton should never be put away damp. That's the recipe for mildew. If you leave a wet load in the wash for more than a day, wash it again before transferring it to the dryer. Canvas tents, awnings, deck chairs and similar outdoor items should be bone dry before storing them.

Permanent creases may form if cotton remains folded too long in the same position. Hang tablecloths over rounded wooden rods or roll up instead of folding. Keep napkins flat until they're ready to use.

Clothes pins and trouser hangers can leave permanent crease marks on waistbands and cuffs. Shoulders may develop unsightly creases from thin wire hangers. For prolonged storage, hang garments on padded hangers and don't overcrowd the closet.

Removal of common stains from cotton

Some stains can be removed effectively with a home remedy. For best results, test the method on a hidden part of the garment. To prevent the stain from spreading, work from the edge toward the center. Thoroughly rinse the cleaned area. When using a commercial spot cleaner, make sure it is safe for cotton.

Berry stains (blackberries, blueberries, raspberries, strawberries): Rub the stain on both sides with bar soap, then cover with a thick mixture of cornstarch and cold water. Rub it in and leave fabric in the sun until the stain disappears. If stain is not gone in three days, repeat. Launder in the usual way.

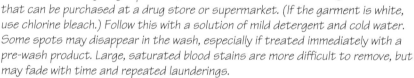

Blood: Act quickly — blood is easiest to remove before it dries. Avoid hot water, which will set the stain. If the stain is still wet, flush with cold water and rub in a bit of laundry detergent or hand soap. If that doesn't work, blot with a dilute solution of hydrogen peroxide, a mild form of bleach that can be purchased at a drug store or supermarket. (If the garment is white, use chlorine bleach.) Follow this with a solution of mild detergent and cold water. Some spots may disappear in the wash, especially if treated immediately with a pre-wash product. Large, saturated blood stains are more difficult to remove, but may fade with time and repeated launderings.

Coffee, tea or chocolate: Treat with warm water and mild detergent, followed by a dilute solution of hydrogen peroxide.

Fruit: Heat and age make fruit stains more difficult to remove, so they need to be treated before the garment is laundered. If treated promptly, fresh fruit stains usually will come out. Sponge them off while they are still wet and scrub with a little detergent. Rinse under hot running water to remove all visible traces of the stain. If that doesn't work, treat with the cornstarch paste described for berry stains.

Ink: There are many different types of ink, so a dry cleaner is recommended. Red ink and permanent ink are types of dye and are hardest to remove.

Lipstick: Water and heat will set the stain and cause it to spread. Rub in vegetable oil and allow to sit for 15 minutes. Sponge in a few drops of ammonia, then wipe it away. Sometimes, hair spray will loosen a lipstick smudge. Spray the stain and allow it to sit for a few minutes, then gently wipe it off.

Oil-based stains: Bicycle grease, food grease, gravy, oil paint, salad dressing and sauces should be removed by a dry cleaner.

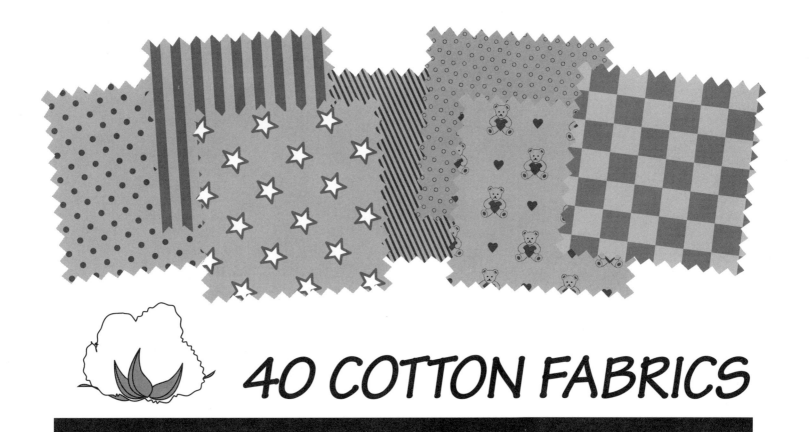

40 COTTON FABRICS

DEFINITIONS & SWATCHES

batik

One of the oldest forms of resist dyeing, originating in Indonesia. Authentic batik is made by coating parts of a plain, undyed fabric with wax so that only the unwaxed areas will take the color. After dyeing, the wax is removed by boiling the fabric or by applying a solvent. The wax-and-dye procedure is repeated for each color in the pattern. The dye seeps through cracks in the wax, creating lovely veins and random streaks of color. Imitation batik is printed by machine, but it does not have random streaks or veins. Batik is usually cotton, but rayon sometimes is used.

Resist dyeing

Resist dyeing is accomplished by treating part of the fabric or yarn with a paste or wax, called resist, to block the dye from penetrating. There are three basic steps: First, the resist is applied to the undyed fabric or yarn. Second, the dye is applied. Third, the resist is removed to reveal a dyed pattern set off by the natural color of the undyed portion. The entire piece can be dyed a second color to produce a two-tone effect, or the resist process can be repeated to form multi-colored designs.

How to use

Batik is a lightweight fabric with a limp, soft or moderately crisp drape. It may be necessary to adjust yardage requirements to accommodate its slightly narrower width. Use to make semi-fitted, loose-fitting or very loose-fitting dresses, blouses, skirts and summer clothing. Machine wash and dry, but beware of shrinkage and dyes that bleed.

Hand-waxed batik has random streaks.

ikat

Handwoven cotton fabrics with distinct, blurred patterns that resemble reflections on water, produced with a variation of tie dyeing. The yarn is tied and dyed before the fabric is woven. The dyes can be applied to lengthwise yarns, crosswise yarns or both. Weavers must be skilled and precise.

resist printing

A pattern is printed on the fabric with a dye-resistant paste. When the fabric is dyed, the unprinted areas become colored. Next, the fabric is washed to remove the paste, producing a white pattern set off by the color. A two-color pattern is produced by adding colored dye to the paste.

tie dyeing

A form of resist dyeing. Portions of the fabric are gathered and tightly knotted with a string or thread. The fabric is then dyed, but the dye doesn't penetrate all the way through the knotted areas, producing irregular spots and streaks. The effect is widely imitated with machine printing.

Sewing rating
- ☒ Easy to sew
- ☐ Moderately easy
- ☐ Average
- ☐ Moderately difficult
- ☐ Extremely difficult

Suggested fit
- ☐ Stretch to fit
- ☐ Close-fitting
- ☐ Fitted
- ☒ Semi-fitted
- ☒ Loose-fitting
- ☒ Very loose-fitting

Suggested styles
- ☒ Pleats ☒ Tucks
  - ☒ pressed
  - ☐ unpressed
- ☒ Gathers
  - ☒ limp ☒ soft
  - ☒ full ☐ lofty
  - ☐ bouffant
- ☒ Elasticized shirring
- ☐ Smocked
- ☐ Tailored
- ☐ Shaped with seams to eliminate bulk
- ☐ Lined
- ☐ Unlined
- ☐ Puffed or bouffant
- ☒ Loose and full
- ☒ Soft and flowing
- ☒ Draped
- ☐ Cut on bias
- ☐ Stretch styling

What to expect
- ☐ Difficult to cut out
- ☐ Fabric has one-way
  - ☐ design ☐ luster
  - ☐ weave ☐ nap
- ☐ Fabric is reversible
- ☐ It looks the same on both sides
- ☐ It stretches easily
- ☐ It will not stretch
- ☒ Fabric tears easily
- ☐ It is difficult to tear
- ☐ Fabric will not tear
- ☐ Pins and needles leave holes, marks
- ☐ It is difficult to ease sleeves and curves
- ☐ It tends to pucker
- ☐ It tends to unravel
- ☐ Inner construction shows from outside
- ☐ Machine eats fabric
- ☐ Skipped stitches
- ☐ Layers feed unevenly
- ☐ Multiple layers are difficult to cut, sew
- ☒ It creases easily
- ☐ Won't hold a crease

Cost per yard
- ☐ Less than $5
- ☒ $5 to $10
- ☒ $10 to $15
- ☐ $15 to $20
- ☐ $20 to $25
- ☐ More than $25

From batik to mud
If your batik fabric is 100 percent cotton, chances are good that the dye isn't especially colorfast. Before you wash the whole piece, test a small sample to see if the colors turn to mud. If they do, you'll need to set the dye or you'll never be able to wash the fabric. Here's one way to do it: Soak the fabric in salt water for 15 minutes, using 3/4 cup of salt per gallon of lukewarm water. If salt crystals form on the fabric, rinse it in clear water and repeat the process, using less salt in the same amount of water. This may not completely set the dye, but it should help and it won't hurt the fabric.

Wearability
- ☒ Durable ☐ Fragile
- ☐ Strong ☐ Weak
- ☐ It is long-wearing
- ☒ It wears evenly
- ☐ It wears out along seams and folds
- ☐ Seams don't hold up under stress
- ☐ Finish wears off
- ☐ Subject to abrasion
- ☒ It resists abrasion
- ☐ Subject to snags
- ☒ It resists snags
- ☐ Subject to runs
- ☐ It tends to pill
- ☐ It tends to shed
- ☐ It produces lint
- ☐ It attracts lint
- ☐ It attract static
- ☐ It tends to cling
- ☒ It holds its shape
- ☐ It loses its shape
- ☐ It stretches out of shape easily
- ☐ It droops, bags
- ☒ It tends to wrinkle
- ☐ It resists wrinkles
- ☐ It crushes easily
- ☐ Water drops leave spots or marks

Suggested care
- ☐ Dry clean only
- ☐ Do not dry clean
- ☐ Dry clean or wash
- ☐ Gently handwash in lukewarm water
- ☐ Roll in a towel to remove moisture
- ☐ Drip dry
- ☐ Lay flat to dry
- ☒ Machine wash
  - ☐ gentle/delicate
  - ☒ regular/normal
- ☒ Machine dry
  - ☐ cool ☒ normal
  - ☒ perm. press
- ☐ Press damp fabric
- ☒ Press dry fabric
- ☐ Dry iron ☒ Steam
- ☐ Iron on wrong side
- ☐ Use a press cloth
- ☐ Use a needleboard
- ☐ Needs no ironing
- ☐ Do not iron
- ☒ Fabric may shrink
- ☒ May bleed or fade
- ☐ Finish washes out

Where to find
- ☐ Any fabric store
- ☒ Major chain store
- ☒ Stores that carry high quality fabric
- ☒ Fabric club
- ☒ Mail order
- ☒ Wholesale supplier

Mercerized cotton

Mercerized cotton is smoother, stronger, more lustrous, easier to dye, more absorbent and more resistant to mildew. The fabric is held under tension to keep it from shrinking while it is soaked in a cold caustic-soda solution. The fiber swells permanently and straightens up, making it easier to apply other finishes. Combed cotton gets the best results and treated fabrics are usually of high quality. The process is also used on linen. It was discovered in 1844 by John Mercer, a calico printer in Lancashire, England.

batiste

Smooth, delicate fabric made with very fine combed cotton yarns in a plain weave. This almost sheer fabric has a soft, limp hand and usually is mercerized to add strength and luster. Batiste is often bleached white or dyed soft pastel colors, but it may also be printed with delicate floral designs. Batiste is similar to lawn, but lawn has more body and is more tightly woven. The first batiste fabric was made of linen and named for Jean Baptiste, a French weaver. Today, linen batiste is hard to find, but the fabric is sometimes made of silk, rayon or wool. Cotton/polyester blends are common and less expensive.

How to use

Batiste has a graceful drape that falls into soft flares. It may be gathered, shirred or smocked into a soft fullness. The fabric is moderately easy to cut and sew, but sheerness may require special finishes to hide seams. Use to make semi-fitted, loose-fitting or very loose-fitting blouses, dresses, lingerie, baby clothes and handkerchiefs. Fabric launders well, but use a gentle cycle.

ATTACH SAMPLE HERE

LENGTHWISE GRAIN

Cotton batiste is smooth and very soft.

---

cambric
Soft, plain-weave cotton or linen cloth that is calendered to give it a slight luster. Finer grades are similar to batiste or lawn, while coarser versions resemble fine muslin.

Nelo batiste
A very fine batiste made in Switzerland.

nainsook
Soft, fine, lightweight fabric made with a plain weave and carded or combed yarns. It has approximately the same number of yarns in both directions of the weave and is similar to batiste, but coarser. Like batiste, it is usually mercerized to add luster. It may be white, pastel or printed.

Swiss batiste
Originally used to describe a fine batiste made in Switzerland. The term is now used more loosely to describe any fine batiste, made in Switzerland or elsewhere of long-staple combed cotton that is mercerized to produce luster. Swiss batiste is usually more expensive than ordinary batiste.

Sewing rating
- [ ] Easy to sew
- [x] Moderately easy
- [ ] Average
- [ ] Moderately difficult
- [ ] Extremely difficult

Suggested fit
- [ ] Stretch to fit
- [ ] Close-fitting
- [ ] Fitted
- [x] Semi-fitted
- [x] Loose-fitting
- [x] Very loose-fitting

Suggested styles
- [ ] Pleats [x] Tucks
  - [x] pressed
  - [ ] unpressed
- [x] Gathers
  - [x] limp [x] soft
  - [ ] full [ ] lofty
  - [ ] bouffant
- [x] Elasticized shirring
- [x] Smocked
- [ ] Tailored
- [ ] Shaped with seams to eliminate bulk
- [ ] Lined
- [ ] Unlined
- [ ] Puffed or bouffant
- [ ] Loose and full
- [x] Soft and flowing
- [ ] Draped
- [ ] Cut on bias
- [ ] Stretch styling

What to expect
- [ ] Difficult to cut out
- [ ] Fabric has one-way
  - [ ] design [ ] luster
  - [ ] weave [ ] nap
- [x] Fabric is reversible
- [x] It looks the same on both sides
- [ ] It stretches easily
- [ ] It will not stretch
- [x] Fabric tears easily
- [ ] It is difficult to tear
- [ ] Fabric will not tear
- [x] Pins and needles leave holes, marks
- [ ] It is difficult to ease sleeves and curves
- [ ] It tends to pucker
- [ ] It tends to unravel
- [x] Inner construction shows from outside
- [x] Machine eats fabric
- [ ] Skipped stitches
- [ ] Layers feed unevenly
- [ ] Multiple layers are difficult to cut, sew
- [x] It creases easily
- [ ] Won't hold a crease

Cost per yard
- [x] Less than $5
- [x] $5 to $10
- [ ] $10 to $15
- [x] $15 to $20
- [x] $20 to $25
- [ ] More than $25

Combed cotton
The finest quality of batiste is made with very thin yarns of combed cotton. All cotton fibers are carded, but only the finest cotton is combed. The combing process removes shorter fibers and impurities from the longer, more desirable fibers. Combed yarns are finer, cleaner, more compact and more even than carded yarns, and fabrics are often tightly woven. While combing is an essential step in the making of very fine cotton yarns, it may also be applied to coarser cotton fibers to improve the quality. Fabrics made from combed cotton are more expensive than ordinary cotton cloths.

Wearability
- [ ] Durable [ ] Fragile
- [ ] Strong [ ] Weak
- [ ] It is long-wearing
- [x] It wears evenly
- [ ] It wears out along seams and folds
- [ ] Seams don't hold up under stress
- [ ] Finish wears off
- [x] Subject to abrasion
- [ ] It resists abrasion
- [ ] Subject to snags
- [x] It resists snags
- [ ] Subject to runs
- [ ] It tends to pill
- [ ] It tends to shed
- [ ] It produces lint
- [ ] It attracts lint
- [ ] It attract static
- [ ] It tends to cling
- [x] It holds its shape
- [ ] It loses its shape
- [ ] It stretches out of shape easily
- [ ] It droops, bags
- [x] It tends to wrinkle
- [ ] It resists wrinkles
- [ ] It crushes easily
- [ ] Water drops leave spots or marks

Suggested care
- [ ] Dry clean only
- [ ] Do not dry clean
- [ ] Dry clean or wash
- [x] Gently handwash in lukewarm water
- [x] Roll in a towel to remove moisture
- [x] Drip dry
- [ ] Lay flat to dry
- [x] Machine wash
  - [x] gentle/delicate
  - [ ] regular/normal
- [x] Machine dry
  - [x] cool [ ] normal
  - [ ] perm. press
- [x] Press damp fabric
- [ ] Press dry fabric
- [x] Dry iron [ ] Steam
- [ ] Iron on wrong side
- [ ] Use a press cloth
- [ ] Use a needleboard
- [ ] Needs no ironing
- [x] Do not iron
- [x] Fabric may shrink
- [ ] May bleed or fade
- [x] Finish washes out

Where to find
- [ ] Any fabric store
- [x] Major chain store
- [x] Stores that carry high quality fabric
- [ ] Fabric club
- [x] Mail order
- [x] Wholesale supplier

broadcloth

Fine, closely woven shirting fabric with very fine crosswise ribs, made from cotton or a cotton/polyester blend. Filling yarns in this plain weave are heavier and have less twist than the lengthwise warp yarns. The best grades are made with combed ply yarns of Pima cotton, while combed singles or carded yarns are used in poorer grades. The fabric usually is mercerized and has a firm hand. It may be bleached white or dyed a solid color, or it may have woven or printed stripes. The term originally referred to any fabric wider than 27-inch narrow cloth. In the 1920s, a lightweight shirting fabric was dubbed "broadcloth" to distinguish it from poplin, which is heavier.

How to use

Broadcloth has a moderately crisp drape and may be pleated, shirred or gathered, but the tight weave may be difficult to ease. It is most often used to make fitted or semi-fitted dress shirts. Dry clean or launder at home. Remove promptly from the dryer to avoid wrinkles. Press with a hot iron.

ATTACH SAMPLE HERE

LENGTHWISE GRAIN

Broadcloth has very fine crosswise ribs.

Pima cotton

Pima cotton flourishes in the irrigated fields of Arizona, New Mexico, Texas and California. The fine, high-grade cotton was developed from a cross between the two best cottons, Egyptian and Sea Island. Pima has extra long, brownish fibers and is used to make fine knitted goods and expensive woven fabrics. When the label says SuPima, it means the product is made from extra long staple Pima cotton grown in the Southwest by one of the 4,000 members of the SuPima Association of America.

---

candy stripes
Stripes of red or pink and white, resembling the stripes on peppermint candy.

chalk stripes
White stripes resembling chalk lines on a darker ground. The width varies, but chalk stripes are usually wider than pin stripes.

double-bar stripes
A stripe pattern made up of two very thin colored stripes paired together against a white or lighter background.

four-bar stripes
Very thin stripes grouped together in sets of four.

pencil stripes
Dark stripes of any width on a lighter background.

pin stripes
Very thin white stripes on a blue or dark background. Very fine pin stripes are sometimes called hairline stripes.

Sewing rating
- ☒ Easy to sew
- ☐ Moderately easy
- ☐ Average
- ☐ Moderately difficult
- ☐ Extremely difficult

Suggested fit
- ☐ Stretch to fit
- ☒ Close-fitting
- ☒ Fitted
- ☒ Semi-fitted
- ☒ Loose-fitting
- ☐ Very loose-fitting

Suggested styles
- ☒ Pleats ☒ Tucks
  - ☒ pressed
  - ☐ unpressed
- ☒ Gathers
  - ☐ limp ☒ soft
  - ☒ full ☐ lofty
  - ☐ bouffant
- ☒ Elasticized shirring
- ☒ Smocked
- ☒ Tailored
- ☐ Shaped with seams to eliminate bulk
- ☐ Lined
- ☐ Unlined
- ☐ Puffed or bouffant
- ☒ Loose and full
- ☐ Soft and flowing
- ☐ Draped
- ☐ Cut on bias
- ☐ Stretch styling

What to expect
- ☐ Difficult to cut out
- ☐ Fabric has one-way
  - ☐ design ☐ luster
  - ☐ weave ☐ nap
- ☒ Fabric is reversible
- ☒ It looks the same on both sides
- ☐ It stretches easily
- ☐ It will not stretch
- ☒ Fabric tears easily
- ☐ It is difficult to tear
- ☐ Fabric will not tear
- ☐ Pins and needles leave holes, marks
- ☒ It is difficult to ease sleeves and curves
- ☐ It tends to pucker
- ☐ It tends to unravel
- ☐ Inner construction shows from outside
- ☐ Machine eats fabric
- ☐ Skipped stitches
- ☐ Layers feed unevenly
- ☐ Multiple layers are difficult to cut, sew
- ☒ It creases easily
- ☐ Won't hold a crease

Cost per yard
- ☒ Less than $5
- ☒ $5 to $10
- ☐ $10 to $15
- ☐ $15 to $20
- ☐ $20 to $25
- ☐ More than $25

Peruvian cotton
The best broadcloths often are made from Pima cotton. One source of the soft, very fine fiber is Peru, where cotton has been cultivated for more than 4,500 years. South America is the origin of the long-staple species, *Gossypium barbadense*, which spawned Egyptian, Sea Island and American Pima cottons. An Egyptian-Sea Island hybrid was mixed with local cottons to produce today's distinct type of Peruvian Pima, which grows along the arid, sunny coast of northern Peru. The plant yields a fine, very white fiber that can be made into silk-like fabrics. Most Peruvian cotton is produced on small, privately owned farms and picked by hand.

Wearability
- ☒ Durable ☐ Fragile
- ☐ Strong ☐ Weak
- ☒ It is long-wearing
- ☒ It wears evenly
- ☐ It wears out along seams and folds
- ☐ Seams don't hold up under stress
- ☐ Finish wears off
- ☐ Subject to abrasion
- ☒ It resists abrasion
- ☐ Subject to snags
- ☒ It resists snags
- ☐ Subject to runs
- ☐ It tends to pill
- ☐ It tends to shed
- ☐ It produces lint
- ☐ It attracts lint
- ☐ It attract static
- ☐ It tends to cling
- ☒ It holds its shape
- ☐ It loses its shape
- ☐ It stretches out of shape easily
- ☐ It droops, bags
- ☒ It tends to wrinkle
- ☐ It resists wrinkles
- ☐ It crushes easily
- ☐ Water drops leave spots or marks

Suggested care
- ☐ Dry clean only
- ☐ Do not dry clean
- ☒ Dry clean or wash
- ☐ Gently handwash in lukewarm water
- ☐ Roll in a towel to remove moisture
- ☐ Drip dry
- ☐ Lay flat to dry
- ☒ Machine wash
  - ☐ gentle/delicate
  - ☒ regular/normal
- ☒ Machine dry
  - ☐ cool ☒ normal
  - ☒ perm. press
- ☐ Press damp fabric
- ☒ Press dry fabric
- ☐ Dry iron ☒ Steam
- ☐ Iron on wrong side
- ☐ Use a press cloth
- ☐ Use a needleboard
- ☐ Needs no ironing
- ☒ Do not iron
- ☒ Fabric may shrink
- ☐ May bleed or fade
- ☐ Finish washes out

Where to find
- ☒ Any fabric store
- ☒ Major chain store
- ☒ Stores that carry high quality fabric
- ☒ Fabric club
- ☒ Mail order
- ☒ Wholesale supplier

Cotton prints

Printing multi-colored designs on cotton is similar to printing on paper – continuous lengths of fabric are fed over a series of engraved rollers. The fabric picks up one color from each roller. A high-speed press can handle 100 to 200 yards a minute and can apply up to 16 colors in a continuous operation. The finished print is dried and baked in an oven to set the dye. Plain weaves are used for most prints because they are economical to produce and have a flat, smooth surface that is ideal for printing.

calico

An inexpensive, firmly woven cotton fabric that is usually made with a plain weave and printed with a small design. Calico was first made in Calicut (Calcutta), India, where fine cotton goods were printed by hand with elaborate wood-block designs and used for decorative purposes. The early fabrics had animals, birds, trees, floral patterns or simple all-over designs. The printing technique was imitated in the United States, where the fabric evolved into a coarse cloth printed by machine, usually with small floral patterns. Today's calico is usually sized and the print doesn't show on the back of the fabric. Other fabrics, such as percale, chintz and printed muslin, are also sold as calico.

How to use

Calico has a moderately limp drape that can be pleated, gathered, smocked or shirred. The fabric is easy to cut and sew. Use to make dresses, children's clothing, quilts and utility linens. Calico is durable and launders well, but the sizing washes out and the fabric tends to shrink the first time it is washed.

ATTACH SAMPLE HERE

LENGTHWISE GRAIN

Cotton calico with a typical floral print.

direct printing

Patterns are printed directly on the fabric in much the same way a newspaper is printed. Each color requires a separate plate, or roller, that is etched with part of the design. The fabric passes between the plates and a master cylinder. Also called roller printing or cylinder printing.

discharge printing

The fabric is first dyed a solid color, then printed with a chemical or bleach to remove the dye, producing a white design against a colored background. A second color may be applied to the design along with the bleach. Unlike many other prints, the finished cloth is uniformly dark on both sides.

screen printing

The pattern is blocked out on a screen of silk or nylon, which is stretched taut over a wooden frame. Dye in the form of colored goo is squeezed through the screen onto the fabric with a squeegee. Each color uses a separate screen. The expensive form of printing is also called silk screening.

Sewing rating
- ☒ Easy to sew
- ☐ Moderately easy
- ☐ Average
- ☐ Moderately difficult
- ☐ Extremely difficult

Suggested fit
- ☐ Stretch to fit
- ☐ Close-fitting
- ☒ Fitted
- ☒ Semi-fitted
- ☒ Loose-fitting
- ☐ Very loose-fitting

Suggested styles
- ☒ Pleats ☒ Tucks
  - ☒ pressed
  - ☐ unpressed
- ☒ Gathers
  - ☐ limp ☒ soft
  - ☒ full ☐ lofty
  - ☐ bouffant
- ☒ Elasticized shirring
- ☒ Smocked
- ☐ Tailored
- ☐ Shaped with seams to eliminate bulk
- ☐ Lined
- ☐ Unlined
- ☒ Puffed or bouffant
- ☒ Loose and full
- ☐ Soft and flowing
- ☐ Draped
- ☐ Cut on bias
- ☐ Stretch styling

What to expect
- ☐ Difficult to cut out
- ☐ Fabric has one-way
  - ☐ design ☐ luster
  - ☐ weave ☐ nap
- ☐ Fabric is reversible
- ☐ It looks the same on both sides
- ☐ It stretches easily
- ☐ It will not stretch
- ☒ Fabric tears easily
- ☐ It is difficult to tear
- ☐ Fabric will not tear
- ☐ Pins and needles leave holes, marks
- ☐ It is difficult to ease sleeves and curves
- ☐ It tends to pucker
- ☐ It tends to unravel
- ☐ Inner construction shows from outside
- ☐ Machine eats fabric
- ☐ Skipped stitches
- ☐ Layers feed unevenly
- ☐ Multiple layers are difficult to cut, sew
- ☒ It creases easily
- ☐ Won't hold a crease

Cost per yard
- ☒ Less than $5
- ☒ $5 to $10
- ☐ $10 to $15
- ☐ $15 to $20
- ☐ $20 to $25
- ☐ More than $25

Upland cotton
Calico and many other fabrics are made from Upland cotton, which makes up the bulk of the U.S. crop. The name was once used to describe any variety of cotton that grew in the U.S. interior, in contrast to the Sea Island varieties, which thrived near the sea. There are many Upland varieties, most with roots in Mexico and Central America. At first, Upland cottons were grown mainly for domestic use. In the late 1700s, the cotton boom inspired development of many improved varieties through hybridization and natural cross-breeding. Today, American Upland is used as the world standard to which all other cotton is compared for grading purposes.

Wearability
- ☒ Durable ☐ Fragile
- ☐ Strong ☐ Weak
- ☐ It is long-wearing
- ☒ It wears evenly
- ☐ It wears out along seams and folds
- ☐ Seams don't hold up under stress
- ☐ Finish wears off
- ☐ Subject to abrasion
- ☐ It resists abrasion
- ☐ Subject to snags
- ☒ It resists snags
- ☐ Subject to runs
- ☐ It tends to pill
- ☐ It tends to shed
- ☐ It produces lint
- ☐ It attracts lint
- ☐ It attract static
- ☐ It tends to cling
- ☒ It holds its shape
- ☐ It loses its shape
- ☐ It stretches out of shape easily
- ☐ It droops, bags
- ☒ It tends to wrinkle
- ☐ It resists wrinkles
- ☐ It crushes easily
- ☐ Water drops leave spots or marks

Suggested care
- ☐ Dry clean only
- ☐ Do not dry clean
- ☐ Dry clean or wash
- ☐ Gently handwash in lukewarm water
- ☐ Roll in a towel to remove moisture
- ☐ Drip dry
- ☐ Lay flat to dry
- ☒ Machine wash
  - ☐ gentle/delicate
  - ☒ regular/normal
- ☒ Machine dry
  - ☐ cool ☒ normal
  - ☐ perm. press
- ☐ Press damp fabric
- ☒ Press dry fabric
- ☐ Dry iron ☒ Steam
- ☐ Iron on wrong side
- ☐ Use a press cloth
- ☐ Use a needleboard
- ☐ Needs no ironing
- ☒ Do not iron
- ☒ Fabric may shrink
- ☐ May bleed or fade
- ☒ Finish washes out

Where to find
- ☒ Any fabric store
- ☒ Major chain store
- ☐ Stores that carry high quality fabric
- ☒ Fabric club
- ☒ Mail order
- ☒ Wholesale supplier

canvas

General term used to describe a variety of stiff, heavy, durable fabrics, originally made of linen or hemp and used for sails. Today, canvas usually is made of cotton, with a tight plain weave and 2- to 14-ply yarns. There are many grades – the fabric may be rough or smooth, unbleached or dyed and softly finished or highly sized. It is often treated with water-repellent or mildew-resistant finishes. The term is used interchangeably with "duck," but duck is usually a rougher, heavier utility cloth, while canvas is smoother and more tightly woven. The term is sometimes used to describe muslin.

How to use

Canvas has a stiff drape that falls in wide cones. The tight weave is difficult to ease and the fabric works best when shaped with seams to eliminate bulk. Multiple layers may be extremely difficult to cut and sew. Use to make tents, awnings, directors' chairs, duffel and tote bags and other soft luggage. Large pieces are difficult to wash by machine because the fabric is so stiff.

ATTACH SAMPLE HERE

LENGTHWISE GRAIN

Canvas is smoother than duck.

Mildew? Phew!

Outdoor fabrics bring out the worst in cotton – it is subject to the growth of mildew, a type of fungus that thrives in warm, damp places. The fungus works quickly to change the cotton's chemical makeup from cellulose to sugar, which it proceeds to eat, eventually causing the fiber to rot. The first sign of trouble is a bad odor, followed by the appearance of dark spots. The odor can be removed with hot water and bleach, but once the spots have taken hold, mildew is nearly impossible to control.

---

awning stripe
Heavy, firm, strong canvas or duck fabric with woven, printed or painted stripes. As a rule, woven stripes are more expensive than painted or printed stripes. Stripes are usually painted or printed on standard fabrics such as army duck, single-filling flat duck and drill.

boatsail drill
Very strong, wind-resistant fabric, made with a plain weave and Egyptian cotton.

marlin
A term used to describe a canvas-like cloth, made of cotton and used to make sails.

sailcloth
General term for fabric used to make sails. Cotton sailcloth is also used for clothing, upholstery and draperies. Polyester and nylon versions are smooth, lightweight, strong, durable and quick-drying, with good resistance to salt water and moisture. Other versions are made of linen and jute.

Sewing rating
- ☐ Easy to sew
- ☐ Moderately easy
- ☐ Average
- ☒ Moderately difficult
- ☐ Extremely difficult

Suggested fit
- ☐ Stretch to fit
- ☒ Close-fitting
- ☒ Fitted
- ☒ Semi-fitted
- ☐ Loose-fitting
- ☐ Very loose-fitting

Suggested styles
- ☐ Pleats ☐ Tucks
  - ☐ pressed
  - ☐ unpressed
- ☐ Gathers
  - ☐ limp ☐ soft
  - ☐ full ☐ lofty
  - ☐ bouffant
- ☐ Elasticized shirring
- ☐ Smocked
- ☐ Tailored
- ☒ Shaped with seams to eliminate bulk
- ☐ Lined
- ☐ Unlined
- ☐ Puffed or bouffant
- ☐ Loose and full
- ☐ Soft and flowing
- ☐ Draped
- ☐ Cut on bias
- ☐ Stretch styling

What to expect
- ☒ Difficult to cut out
- ☐ Fabric has one-way
  - ☐ design ☐ luster
  - ☐ weave ☐ nap
- ☒ Fabric is reversible
- ☒ It looks the same on both sides
- ☐ It stretches easily
- ☒ It will not stretch
- ☐ Fabric tears easily
- ☒ It is difficult to tear
- ☒ Fabric will not tear
- ☒ Pins and needles leave holes, marks
- ☒ It is difficult to ease sleeves and curves
- ☐ It tends to pucker
- ☐ It tends to unravel
- ☐ Inner construction shows from outside
- ☐ Machine eats fabric
- ☒ Skipped stitches
- ☒ Layers feed unevenly
- ☒ Multiple layers are difficult to cut, sew
- ☐ It creases easily
- ☒ Won't hold a crease

Cost per yard
- ☒ Less than $5
- ☒ $5 to $10
- ☒ $10 to $15
- ☐ $15 to $20
- ☐ $20 to $25
- ☐ More than $25

Preventing mildew
Canvas and other cotton fabrics should be stored clean and dry to stop mildew from making an appearance.
During periods of high humidity, tents, awnings and other outdoor items should be aired in the sun frequently. It helps when the fabric has been treated with a special finish to curb the growth of mold, mildew and bacteria, control the spread of germs and reduce the development of unpleasant odors. Water repellents and flame retardants also prevent the growth of mildew. Special sprays are available for home use, but they are not very durable and must be re-applied after each wash.

Wearability
- ☒ Durable ☐ Fragile
- ☒ Strong ☐ Weak
- ☒ It is long-wearing
- ☒ It wears evenly
- ☐ It wears out along seams and folds
- ☐ Seams don't hold up under stress
- ☐ Finish wears off
- ☐ Subject to abrasion
- ☒ It resists abrasion
- ☐ Subject to snags
- ☒ It resists snags
- ☐ Subject to runs
- ☐ It tends to pill
- ☐ It tends to shed
- ☐ It produces lint
- ☐ It attracts lint
- ☐ It attract static
- ☐ It tends to cling
- ☒ It holds its shape
- ☐ It loses its shape
- ☐ It stretches out of shape easily
- ☐ It droops, bags
- ☐ It tends to wrinkle
- ☒ It resists wrinkles
- ☐ It crushes easily
- ☐ Water drops leave spots or marks

Suggested care
- ☐ Dry clean only
- ☐ Do not dry clean
- ☐ Dry clean or wash
- ☐ Gently handwash in lukewarm water
- ☐ Roll in a towel to remove moisture
- ☒ Drip dry
- ☐ Lay flat to dry
- ☒ Machine wash
  - ☐ gentle/delicate
  - ☒ regular/normal
- ☒ Machine dry
  - ☐ cool ☒ normal
  - ☐ perm. press
- ☐ Press damp fabric
- ☐ Press dry fabric
- ☐ Dry iron ☐ Steam
- ☐ Iron on wrong side
- ☐ Use a press cloth
- ☐ Use a needleboard
- ☒ Needs no ironing
- ☐ Do not iron
- ☒ Fabric may shrink
- ☐ May bleed or fade
- ☒ Finish washes out

Where to find
- ☒ Any fabric store
- ☒ Major chain store
- ☒ Stores that carry high quality fabric
- ☐ Fabric club
- ☒ Mail order
- ☒ Wholesale supplier

chambray

Soft, comfortable shirt fabric made with a plain weave, colored warp yarns and white fillings, giving it a weathered look. The flat, smooth fabric may be cotton or a blend of cotton and polyester. It is usually a solid pastel color, but some fabrics have woven checks or stripes, and it is sometimes embellished with machine embroidery. Other versions are printed to look like chambray. The fabric may be treated to resist wrinkles and shrinkage. It varies in weight — heavier versions may be marked by weight and sold along side denim fabrics. Named from Cambrai, France, where it was first made.

How to use

Chambray has a comfortable drape that falls into moderately soft flares. It is easy to cut and sew and may be pleated, gathered or shirred into a soft fullness. Use to make semi-fitted, loose or very loose dresses, skirts, casual work shirts and children's clothes. Chambray wears well, launders easily and resists wrinkles, especially when made of a cotton/polyester blend.

Chambray is often pale blue or pink.

End-and-end

Chambray and other shirting fabrics often are woven with an "end-and-end" pattern. "End" is textile industry slang for the warp, or lengthwise, yarns of a fabric. End-and-end fabric is made by alternating white ends with colored ends. The filling yarns (also called "picks") are all one color, usually white. The finished fabric has lengthwise colored stripes on a white background, or white stripes on a colored background. The size of the stripe varies depending on how the yarns are grouped.

fancy chambray
A fine dress chambray made with combed single yarns and a clipped spot or small dobby design on a chambray background.

oxford chambray
An oxford cloth woven like chambray, with colored warp yarns and white fillings.

shirting chambray
A fine shirting made of combed single yarns and a high thread count of 80x76 or more. Fabric often has white selvages and is sometimes made with an end-and-end effect, alternating one end of color and one end of white. It is usually bleached and may be mercerized and pre-shrunk.

workshirt chambray
A popular chambray fabric made of carded cotton in a number of different weights and constructions. The colored warp yarns are heavier than the filling yarns, which are usually gray or white, and fabric has many more warps than fillings per inch. Used for men's shirts and children's play clothes.

Sewing rating
- ☒ Easy to sew
- ☐ Moderately easy
- ☐ Average
- ☐ Moderately difficult
- ☐ Extremely difficult

Suggested fit
- ☐ Stretch to fit
- ☐ Close-fitting
- ☒ Fitted
- ☒ Semi-fitted
- ☒ Loose-fitting
- ☐ Very loose-fitting

Suggested styles
- ☒ Pleats ☒ Tucks
  - ☒ pressed
  - ☐ unpressed
- ☒ Gathers
  - ☐ limp ☒ soft
  - ☒ full ☐ lofty
  - ☐ bouffant
- ☒ Elasticized shirring
- ☐ Smocked
- ☒ Tailored
- ☐ Shaped with seams to eliminate bulk
- ☐ Lined
- ☐ Unlined
- ☐ Puffed or bouffant
- ☒ Loose and full
- ☐ Soft and flowing
- ☐ Draped
- ☐ Cut on bias
- ☐ Stretch styling

What to expect
- ☐ Difficult to cut out
- ☐ Fabric has one-way
  - ☐ design ☐ luster
  - ☐ weave ☐ nap
- ☒ Fabric is reversible
- ☒ It looks the same on both sides
- ☐ It stretches easily
- ☐ It will not stretch
- ☒ Fabric tears easily
- ☐ It is difficult to tear
- ☐ Fabric will not tear
- ☐ Pins and needles leave holes, marks
- ☒ It is difficult to ease sleeves and curves
- ☐ It tends to pucker
- ☐ It tends to unravel
- ☐ Inner construction shows from outside
- ☐ Machine eats fabric
- ☐ Skipped stitches
- ☐ Layers feed unevenly
- ☐ Multiple layers are difficult to cut, sew
- ☒ It creases easily
- ☐ Won't hold a crease

Cost per yard
- ☒ Less than $5
- ☒ $5 to $10
- ☐ $10 to $15
- ☐ $15 to $20
- ☐ $20 to $25
- ☐ More than $25

The plain weave
Most fabrics are made with one of the three basic weaves — plain, twill and satin. The plain weave, also called a tabby weave, is the simplest and most common weave, used in nearly 80 percent of all woven fabrics. Each lengthwise warp yarn passes over-under-over-under each crosswise filling yarn, forming a checkerboard pattern. Fabrics are smooth and flat. They print beautifully and wear evenly, but wrinkle badly and shrink. Fabrics tear easily and are not as strong or firm as most twills. Twill and satin fabrics are more absorbent and have a more fluid drape.

Wearability
- ☒ Durable ☐ Fragile
- ☐ Strong ☐ Weak
- ☒ It is long-wearing
- ☒ It wears evenly
- ☐ It wears out along seams and folds
- ☐ Seams don't hold up under stress
- ☐ Finish wears off
- ☐ Subject to abrasion
- ☒ It resists abrasion
- ☐ Subject to snags
- ☒ It resists snags
- ☐ Subject to runs
- ☐ It tends to pill
- ☐ It tends to shed
- ☐ It produces lint
- ☐ It attracts lint
- ☐ It attract static
- ☐ It tends to cling
- ☒ It holds its shape
- ☐ It loses its shape
- ☐ It stretches out of shape easily
- ☐ It droops, bags
- ☐ It tends to wrinkle
- ☒ It resists wrinkles
- ☐ It crushes easily
- ☐ Water drops leave spots or marks

Suggested care
- ☐ Dry clean only
- ☐ Do not dry clean
- ☐ Dry clean or wash
- ☐ Gently handwash in lukewarm water
- ☐ Roll in a towel to remove moisture
- ☐ Drip dry
- ☐ Lay flat to dry
- ☒ Machine wash
  - ☐ gentle/delicate
  - ☒ regular/normal
- ☒ Machine dry
  - ☐ cool ☒ normal
  - ☒ perm. press
- ☐ Press damp fabric
- ☒ Press dry fabric
- ☐ Dry iron ☒ Steam
- ☐ Iron on wrong side
- ☐ Use a press cloth
- ☐ Use a needleboard
- ☐ Needs no ironing
- ☐ Do not iron
- ☒ Fabric may shrink
- ☐ May bleed or fade
- ☒ Finish washes out

Where to find
- ☒ Any fabric store
- ☒ Major chain store
- ☒ Stores that carry high quality fabric
- ☒ Fabric club
- ☒ Mail order
- ☒ Wholesale supplier

The lint factor

Lint is a term used to describe all those pesky short fibers and soft pieces of fuzz that come loose from some fabrics, usually in the wash, and stick to others. Fabrics like chenille and corduroy are big lint producers because they are made with cut yarns, enabling the short cotton fibers to unravel easily. Lint is attracted to fabrics that collect static, a common characteristic of synthetics. To avoid lint buildup, fabrics that produce a lot of lint should not be washed or dried with fabrics that collect static.

chenille

Any fabric woven with chenille yarn, a soft fluffy yarn with pile protruding on all sides like the tufts on a caterpillar's back. The expensive yarn usually is inserted as an extra filling into a conventionally woven or knitted background, producing a thick, firm fabric with soft velvety pile on both sides. When made of cotton, chenille is super absorbent. It is sometimes made of rayon, acrylic or wool. Chenille is used to make towels, bath mats, upholstery and drapery fabrics, rugs, fringes and tassels. Chenille is French for caterpillar.

How to use

Chenille has a soft, comfortable drape and may be gathered into a thick heavy fullness or shaped with seams to eliminate bulk. Like most pile fabrics, chenille tends to shed and the pile eventually wears off. Chenille is a popular bathrobe fabric, but it may also be used to make bedspreads, jackets and casual tops. Chenille launders well, but it dries slowly and produces a lot of lint. The fabric fluffs up in the dryer and does not need to be ironed.

ATTACH SAMPLE HERE

LENGTHWISE GRAIN

Chenille is thick, fuzzy and absorbent.

---

bouclé yarn
A three-ply yarn with small, tight loops protruding at widely spaced intervals.

brushed yarn
A yarn made with short staple fibers that are brushed to the surface to produce a soft, bulky effect.

candlewick fabric
A tufted pile fabric made from unbleached muslin decorated with candlewick yarns, producing a pile similar to that of chenille.

candlewick yarn
A heavy, plied cotton yarn used to make wicks of candles and to decorate fabrics.

corkscrew yarn
A two-ply yarn made by twisting a thin, tightly twisted yarn around a thick, loosely twisted yarn.

spiral yarn
A two-ply yarn made by twisting a soft, thick yarn around a fine yarn.

Sewing rating
- [ ] Easy to sew
- [ ] Moderately easy
- [ ] Average
- [x] Moderately difficult
- [ ] Extremely difficult

Suggested fit
- [ ] Stretch to fit
- [ ] Close-fitting
- [ ] Fitted
- [x] Semi-fitted
- [x] Loose-fitting
- [x] Very loose-fitting

Suggested styles
- [ ] Pleats [x] Tucks
  - [ ] pressed
  - [x] unpressed
- [x] Gathers
  - [ ] limp [ ] soft
  - [x] full [ ] lofty
  - [ ] bouffant
- [ ] Elasticized shirring
- [ ] Smocked
- [ ] Tailored
- [x] Shaped with seams to eliminate bulk
- [ ] Lined
- [ ] Unlined
- [ ] Puffed or bouffant
- [x] Loose and full
- [ ] Soft and flowing
- [ ] Draped
- [ ] Cut on bias
- [ ] Stretch styling

What to expect
- [x] Difficult to cut out
- [ ] Fabric has one-way
  - [ ] design [ ] luster
  - [ ] weave [ ] nap
- [x] Fabric is reversible
- [x] It looks the same on both sides
- [ ] It stretches easily
- [ ] It will not stretch
- [ ] Fabric tears easily
- [ ] It is difficult to tear
- [x] Fabric will not tear
- [ ] Pins and needles leave holes, marks
- [ ] It is difficult to ease sleeves and curves
- [ ] It tends to pucker
- [ ] It tends to unravel
- [ ] Inner construction shows from outside
- [ ] Machine eats fabric
- [x] Skipped stitches
- [x] Layers feed unevenly
- [x] Multiple layers are difficult to cut, sew
- [ ] It creases easily
- [x] Won't hold a crease

Cost per yard
- [ ] Less than $5
- [ ] $5 to $10
- [ ] $10 to $15
- [x] $15 to $20
- [x] $20 to $25
- [x] More than $25

Chenille yarns
A chenille yarn looks like a limp pipe cleaner – without the wire to make it stiff. Chenille yarns are made by weaving a fabric with very tightly spaced filling yarns and warp yarns arranged in groups of four. When finished, the fabric is cut in the lengthwise direction between each group of warp yarns. Each piece is twisted, producing a cord-like yarn with pile protruding on all sides. The warp yarns form the cord, the cut filling yarns form the pile and the twist holds it all together. It is an expensive way to make yarn, so chenille fabrics tend to be costly.

Wearability
- [x] Durable [ ] Fragile
- [x] Strong [ ] Weak
- [x] It is long-wearing
- [ ] It wears evenly
- [x] It wears out along seams and folds
- [ ] Seams don't hold up under stress
- [ ] Finish wears off
- [x] Subject to abrasion
- [ ] It resists abrasion
- [ ] Subject to snags
- [ ] It resists snags
- [ ] Subject to runs
- [ ] It tends to pill
- [x] It tends to shed
- [x] It produces lint
- [ ] It attracts lint
- [ ] It attract static
- [ ] It tends to cling
- [x] It holds its shape
- [ ] It loses its shape
- [ ] It stretches out of shape easily
- [ ] It droops, bags
- [ ] It tends to wrinkle
- [x] It resists wrinkles
- [ ] It crushes easily
- [ ] Water drops leave spots or marks

Suggested care
- [ ] Dry clean only
- [ ] Do not dry clean
- [ ] Dry clean or wash
- [ ] Gently handwash in lukewarm water
- [ ] Roll in a towel to remove moisture
- [ ] Drip dry
- [ ] Lay flat to dry
- [x] Machine wash
  - [ ] gentle/delicate
  - [x] regular/normal
- [x] Machine dry
  - [ ] cool [x] normal
  - [ ] perm. press
- [ ] Press damp fabric
- [ ] Press dry fabric
- [ ] Dry iron [ ] Steam
- [ ] Iron on wrong side
- [ ] Use a press cloth
- [ ] Use a needleboard
- [x] Needs no ironing
- [ ] Do not iron
- [x] Fabric may shrink
- [ ] May bleed or fade
- [ ] Finish washes out

Where to find
- [ ] Any fabric store
- [ ] Major chain store
- [x] Stores that carry high quality fabric
- [ ] Fabric club
- [x] Mail order
- [x] Wholesale supplier

chino

Durable cotton fabric with a noticeable diagonal twill line on the front and a plain back. Usually, chino is made of combed, two-ply cotton yarns and mercerized to increase its strength and luster. It has a slight sheen on the face and a dull back, and is frequently tan (khaki), but it is also available in other colors. The firm, compact fabric varies in weight from medium to heavy and usually is treated with a special finish to reduce wrinkles and shrinkage. It is favored for military uniforms because it withstands all sorts of abuse.

How to use

Chino has a moderately stiff drape that falls into wide flares and retains the shape of the garment. It is easy to cut and sew, but the tight weave may be difficult to ease. The fabric wears well and resists snags and abrasions. Use to make fitted, semi-fitted or loose-fitting men's slacks, sportswear, work clothes and uniforms. The fabric holds up indefinitely to numerous washings. Machine wash and tumble dry.

Chino has a slight luster on the face.

That khaki color

Back in the 1800s, before chino was chino, it was used to make uniforms for British troops in India, who waged a war against dust. Soldiers dyed their white uniforms with coffee and curry powder and called them khaki – Hindu for the color of dust. The fabric was first made in China and later in England, where it was dyed a similar color, sent to China and purchased by the U.S. Army to make uniforms for troops in the Philippines. It was dubbed "chino," apparently because it came from China.

Sanforized®
Registered trademark of Cluett, Peabody and Co., Inc., applied to cotton fabrics and cotton/synthetic blends that have been mechanically treated to reduce shrinkage to 1 percent or less in either direction. The label means the quality has been checked by the trademark's owners.

Sanforized-Plus®
To carry this registered trademark, fabrics must meet rigid standards for strength, minimal shrinkage and crease recovery.

Sanforized-Plus-2®
Durable-press trademark that is licensed to clothing manufacturers.

Sanfor-Set®
A registered trademark of Cluett, Peabody & Co., Inc., for an easy-care, durable-press finish. To qualify for this label, fabric must meet rigid standards for shrinkage control. Fabrics also must have a soft and supple hand, a smooth appearance and better resistance to wear, among other qualities.

Sewing rating
- [ ] Easy to sew
- [x] Moderately easy
- [ ] Average
- [ ] Moderately difficult
- [ ] Extremely difficult

Suggested fit
- [ ] Stretch to fit
- [ ] Close-fitting
- [x] Fitted
- [x] Semi-fitted
- [x] Loose-fitting
- [ ] Very loose-fitting

Suggested styles
- [x] Pleats [x] Tucks
  - [ ] pressed
  - [x] unpressed
- [ ] Gathers
  - [ ] limp [ ] soft
  - [ ] full [ ] lofty
  - [ ] bouffant
- [ ] Elasticized shirring
- [ ] Smocked
- [x] Tailored
- [x] Shaped with seams to eliminate bulk
- [ ] Lined
- [ ] Unlined
- [ ] Puffed or bouffant
- [x] Loose and full
- [ ] Soft and flowing
- [x] Draped
- [ ] Cut on bias
- [ ] Stretch styling

What to expect
- [ ] Difficult to cut out
- [x] Fabric has one-way
  - [ ] design [x] luster
  - [x] weave [ ] nap
- [ ] Fabric is reversible
- [ ] It looks the same on both sides
- [ ] It stretches easily
- [x] It will not stretch
- [ ] Fabric tears easily
- [x] It is difficult to tear
- [ ] Fabric will not tear
- [ ] Pins and needles leave holes, marks
- [x] It is difficult to ease sleeves and curves
- [ ] It tends to pucker
- [ ] It tends to unravel
- [ ] Inner construction shows from outside
- [ ] Machine eats fabric
- [ ] Skipped stitches
- [ ] Layers feed unevenly
- [ ] Multiple layers are difficult to cut, sew
- [ ] It creases easily
- [x] Won't hold a crease

Cost per yard
- [ ] Less than $5
- [ ] $5 to $10
- [x] $10 to $15
- [x] $15 to $20
- [ ] $20 to $25
- [ ] More than $25

Shrinkage busters
The relaxation shrinkage that plagues many cotton fabrics can be minimized by treating the fabric with a special finish. One method is to compress wet fabric in the lengthwise direction by overfeeding it onto a large roller covered by felt or a rubber blanket. It is then dried in this form. Before the process, each fabric is tested to determine its potential shrinkage and then is forced to shrink by that amount. Such fabrics are called "compacted" or "compressed." Sanforized® is a trademarked name for such a treatment. The finish is more permanent than a resin finish, which gradually washes out and loses its effectiveness.

STOP SHRINKAGE

Wearability
- [x] Durable [ ] Fragile
- [x] Strong [ ] Weak
- [x] It is long-wearing
- [ ] It wears evenly
- [x] It wears out along seams and folds
- [ ] Seams don't hold up under stress
- [ ] Finish wears off
- [ ] Subject to abrasion
- [x] It resists abrasion
- [ ] Subject to snags
- [x] It resists snags
- [ ] Subject to runs
- [ ] It tends to pill
- [ ] It tends to shed
- [ ] It produces lint
- [ ] It attracts lint
- [ ] It attract static
- [ ] It tends to cling
- [x] It holds its shape
- [ ] It loses its shape
- [ ] It stretches out of shape easily
- [ ] It droops, bags
- [ ] It tends to wrinkle
- [x] It resists wrinkles
- [ ] It crushes easily
- [x] Water drops leave spots or marks

Suggested care
- [ ] Dry clean only
- [ ] Do not dry clean
- [ ] Dry clean or wash
- [ ] Gently handwash in lukewarm water
- [ ] Roll in a towel to remove moisture
- [ ] Drip dry
- [ ] Lay flat to dry
- [x] Machine wash
  - [ ] gentle/delicate
  - [x] regular/normal
- [x] Machine dry
  - [ ] cool [x] normal
  - [x] perm. press
- [ ] Press damp fabric
- [x] Press dry fabric
- [ ] Dry iron [x] Steam
- [ ] Iron on wrong side
- [ ] Use a press cloth
- [ ] Use a needleboard
- [ ] Needs no ironing
- [ ] Do not iron
- [x] Fabric may shrink
- [ ] May bleed or fade
- [ ] Finish washes out

Where to find
- [ ] Any fabric store
- [x] Major chain store
- [x] Stores that carry high quality fabric
- [ ] Fabric club
- [x] Mail order
- [x] Wholesale supplier

chintz

A closely woven, plain-weave cotton fabric with a very lustrous, glazed finish. Chintz is usually printed with large, brightly colored floral patterns, but it may also be dyed a solid color or printed with stripes, dots or other geometric designs. The crisp, firm fabric is made with fine, tightly twisted warp yarns and coarser, low-twist filling yarns. It has a smooth, shiny face and a dull back. Some versions are given an additional finish, such as Scotchgard®, to make them repel water and soil. The popular fabric was first made in India. Chintes is the plural of the Hindu word chit, meaning spotted or variegated.

How to use

Chintz has a crisp drape that falls into stiff flares. It may be gathered into a bouffant fullness, pleated into sharp folds or shaped with seams to reduce bulk. Chintz resists snags and abrasion, but the shine tends to grow dull with wear. Use to make draperies, slip covers, dresses and jackets. Some chintz can be washed, but it usually needs to be dry cleaned to protect the finish.

Large floral patterns are typical of chintz.

Glazed finishes

The first chintz was a calico fabric that was painted by hand and slicked with starch, in India. The starch washed out, so the fabric was used for decorative purposes rather than clothing. Today's modern glazes are more durable, produced by treating the cloth with a chemical finish that withstands many washes. Some lower grades of chintz are still given a wax or starch glaze, applied by running the fabric through a series of hot rollers. These glazes tend to wash out, so fabrics must be dry cleaned.

---

chintz-finished print

Cotton or cotton/polyester blend that is printed with a chintz-like floral pattern and given a chintz-like finish. Fabric is lighter in weight, not as stiff and not as lustrous as authentic chintz. It is intended for clothing, rather than for decorative purposes, and is less expensive than authentic chintz.

cretonne

A decorative fabric similar to chintz, but treated with a dull, rather than lustrous, finish. It is usually printed with brightly colored floral designs on a natural-colored background. It may be all cotton, all linen or blended with synthetics. Used to make draperies and slipcovers.

polished cotton

Any cotton or cotton/polyester fabric that has been treated with a lustrous finish similar to chintz. Fabric is usually lighter in weight than chintz and is not as lustrous or stiff. The finish may not be permanent. Polished Apple® is a registered trademark for a type of polished cotton.

Sewing rating
- [ ] Easy to sew
- [x] Moderately easy
- [ ] Average
- [ ] Moderately difficult
- [ ] Extremely difficult

Suggested fit
- [ ] Stretch to fit
- [x] Close-fitting
- [x] Fitted
- [x] Semi-fitted
- [x] Loose-fitting
- [ ] Very loose-fitting

Suggested styles
- [x] Pleats [x] Tucks
  - [x] pressed
  - [ ] unpressed
- [x] Gathers
  - [ ] limp [ ] soft
  - [ ] full [x] lofty
  - [x] bouffant
- [x] Elasticized shirring
- [ ] Smocked
- [x] Tailored
- [x] Shaped with seams to eliminate bulk
- [x] Lined
- [x] Unlined
- [x] Puffed or bouffant
- [ ] Loose and full
- [ ] Soft and flowing
- [ ] Draped
- [ ] Cut on bias
- [ ] Stretch styling

What to expect
- [ ] Difficult to cut out
- [x] Fabric has one-way
  - [ ] design [x] luster
  - [ ] weave [ ] nap
- [ ] Fabric is reversible
- [ ] It looks the same on both sides
- [ ] It stretches easily
- [x] It will not stretch
- [ ] Fabric tears easily
- [x] It is difficult to tear
- [ ] Fabric will not tear
- [ ] Pins and needles leave holes, marks
- [x] It is difficult to ease sleeves and curves
- [ ] It tends to pucker
- [ ] It tends to unravel
- [ ] Inner construction shows from outside
- [ ] Machine eats fabric
- [ ] Skipped stitches
- [ ] Layers feed unevenly
- [ ] Multiple layers are difficult to cut, sew
- [x] It creases easily
- [ ] Won't hold a crease

Cost per yard
- [ ] Less than $5
- [x] $5 to $10
- [x] $10 to $15
- [ ] $15 to $20
- [ ] $20 to $25
- [ ] More than $25

Stain-repellent finishes
Cotton is easily cleaned because it absorbs water easily, which in turn helps it to release soil. This ability is diminished when the fiber is given a glazed finish like the one applied to chintz. The problem is complicated by the fact that glazed finishes tend to wash out, so chintz is often given a second, stain-repellent finish to protect the glazed fabric and reduce washings. There are two types: One coats the fiber to prevent soil penetration; the other increases the treated fiber's ability to absorb water and thus, release soil. Scotch Release®, Scotchgard®, Visa® and Zepel® are all registered trademarks for stain-repellent finishes.

Wearability
- [x] Durable [ ] Fragile
- [ ] Strong [ ] Weak
- [x] It is long-wearing
- [ ] It wears evenly
- [x] It wears out along seams and folds
- [ ] Seams don't hold up under stress
- [x] Finish wears off
- [ ] Subject to abrasion
- [x] It resists abrasion
- [ ] Subject to snags
- [x] It resists snags
- [ ] Subject to runs
- [ ] It tends to pill
- [ ] It tends to shed
- [ ] It produces lint
- [ ] It attracts lint
- [ ] It attract static
- [ ] It tends to cling
- [x] It holds its shape
- [ ] It loses its shape
- [ ] It stretches out of shape easily
- [ ] It droops, bags
- [x] It tends to wrinkle
- [ ] It resists wrinkles
- [ ] It crushes easily
- [x] Water drops leave spots or marks

Suggested care
- [ ] Dry clean only
- [ ] Do not dry clean
- [x] Dry clean or wash
- [ ] Gently handwash in lukewarm water
- [ ] Roll in a towel to remove moisture
- [ ] Drip dry
- [ ] Lay flat to dry
- [x] Machine wash
  - [ ] gentle/delicate
  - [x] regular/normal
- [x] Machine dry
  - [ ] cool [x] normal
  - [x] perm. press
- [ ] Press damp fabric
- [x] Press dry fabric
- [x] Dry iron [ ] Steam
- [x] Iron on wrong side
- [ ] Use a press cloth
- [ ] Use a needleboard
- [ ] Needs no ironing
- [ ] Do not iron
- [x] Fabric may shrink
- [ ] May bleed or fade
- [x] Finish washes out

Where to find
- [x] Any fabric store
- [x] Major chain store
- [x] Stores that carry high quality fabric
- [x] Fabric club
- [x] Mail order
- [x] Wholesale supplier

corduroy

A thick, rugged pile fabric made of cotton or a cotton/polyester blend with a plain or twill weave. Corduroy has a soft distinctive pile that forms lengthwise ribs, called wales, produced by an extra set of filling yarns that are cut and brushed. The wales vary in size and height, from narrow to wide and from flat to plush. The weight varies from soft shirting fabrics to rugged upholstery goods. Heavier versions tend to have wider wales. Corduroy is a firm, sturdy fabric. It may be solid in color or printed with plaid, floral or other patterns.

How to use

Corduroy has a moderately stiff drape that falls into soft, wide flares. It may be lightly tucked or gathered into a lofty fullness or shaped with seams to eliminate bulk. The napped fabric requires a one-way cutting layout. Heavier versions may be especially difficult to cut and sew. Corduroy is durable, but the pile tends to wear off. Use for sportswear, coats, jackets, shirts, slacks, skirts and children's clothing. Dryclean or launder at home.

Corduroy's pile forms soft lengthwise wales.

A royal past

Corduroy got its name in the 1600s, when the popular fabric was worn by servants of the French king and dubbed "cord du roi," which is French for "cord of the king." In the mid-1700s, a British textile importer used the royal link to promote his sales and the name stuck. The fabric is no longer associated with royalty, but it can still be quite luxurious. Fine corduroy is made of combed cotton and is softer, more lustrous and more expensive. The most luxurious corduroy is made with silk pile.

---

fine-wale corduroy
Corduroy with more than 21 wales per inch.

Genoa cord
Corduroy woven with a twill background.

jacquard-cut corduroy
Pile is cut to resemble a woven jacquard.

jumbo-wale corduroy
Corduroy with three to 10 wales per inch.

midwale corduroy
Corduroy with 11 to 15 wales per inch.

pinwale corduroy
Corduroy with 16 to 21 wales per inch.

tabby-back corduroy
Corduroy woven with a plain-weave backing.

thick-and-thin wale corduroy
Fabric has alternating thick and thin wales.

wide-wale corduroy
Another name for jumbo-wale corduroy.

Sewing rating
- ☐ Easy to sew
- ☐ Moderately easy
- ☐ Average
- ☒ Moderately difficult
- ☐ Extremely difficult

Suggested fit
- ☐ Stretch to fit
- ☐ Close-fitting
- ☒ Fitted
- ☒ Semi-fitted
- ☒ Loose-fitting
- ☐ Very loose-fitting

Suggested styles
- ☐ Pleats ☒ Tucks
  - ☐ pressed
  - ☒ unpressed
- ☒ Gathers
  - ☐ limp ☐ soft
  - ☒ full ☒ lofty
  - ☐ bouffant
- ☐ Elasticized shirring
- ☐ Smocked
- ☒ Tailored
- ☒ Shaped with seams to eliminate bulk
- ☒ Lined
- ☒ Unlined
- ☐ Puffed or bouffant
- ☒ Loose and full
- ☐ Soft and flowing
- ☐ Draped
- ☐ Cut on bias
- ☐ Stretch styling

What to expect
- ☒ Difficult to cut out
- ☒ Fabric has one-way
  - ☐ design ☐ luster
  - ☐ weave ☒ nap
- ☐ Fabric is reversible
- ☐ It looks the same on both sides
- ☐ It stretches easily
- ☒ It will not stretch
- ☐ Fabric tears easily
- ☒ It is difficult to tear
- ☒ Fabric will not tear
- ☐ Pins and needles leave holes, marks
- ☒ It is difficult to ease sleeves and curves
- ☐ It tends to pucker
- ☐ It tends to unravel
- ☐ Inner construction shows from outside
- ☐ Machine eats fabric
- ☒ Skipped stitches
- ☒ Layers feed unevenly
- ☒ Multiple layers are difficult to cut, sew
- ☐ It creases easily
- ☒ Won't hold a crease

Cost per yard
- ☐ Less than $5
- ☒ $5 to $10
- ☒ $10 to $15
- ☒ $15 to $20
- ☐ $20 to $25
- ☐ More than $25

How corduroy's wales are formed
Corduroy's wales are formed by weaving an extra set of crosswise filling yarns into a background fabric made with a plain or twill weave. The extra yarns "float" over three or more warp yarns, under one or two, then over three and so forth. After weaving, the floats are cut in the middle and the fibers spring upward. Later, they are brushed to form the ridges, better known as wales. The ridges are rounded, with the longest fibers in the center, formed by longer floats, and the shortest fibers on each side, formed by shorter floats. The weave can be varied to produce wales of varying heights and thicknesses in the same fabric.

Wearability
- ☒ Durable ☐ Fragile
- ☒ Strong ☐ Weak
- ☒ It is long-wearing
- ☐ It wears evenly
- ☒ It wears out along seams and folds
- ☐ Seams don't hold up under stress
- ☐ Finish wears off
- ☒ Subject to abrasion
- ☐ It resists abrasion
- ☐ Subject to snags
- ☒ It resists snags
- ☐ Subject to runs
- ☐ It tends to pill
- ☒ It tends to shed
- ☒ It produces lint
- ☐ It attracts lint
- ☐ It attract static
- ☐ It tends to cling
- ☒ It holds its shape
- ☐ It loses its shape
- ☐ It stretches out of shape easily
- ☐ It droops, bags
- ☐ It tends to wrinkle
- ☒ It resists wrinkles
- ☒ It crushes easily
- ☒ Water drops leave spots or marks

Suggested care
- ☐ Dry clean only
- ☐ Do not dry clean
- ☒ Dry clean or wash
- ☐ Gently handwash in lukewarm water
- ☐ Roll in a towel to remove moisture
- ☐ Drip dry
- ☐ Lay flat to dry
- ☒ Machine wash
  - ☐ gentle/delicate
  - ☒ regular/normal
- ☒ Machine dry
  - ☐ cool ☒ normal
  - ☐ perm. press
- ☐ Press damp fabric
- ☒ Press dry fabric
- ☐ Dry iron ☒ Steam
- ☒ Iron on wrong side
- ☐ Use a press cloth
- ☐ Use a needleboard
- ☒ Needs no ironing
- ☐ Do not iron
- ☒ Fabric may shrink
- ☐ May bleed or fade
- ☐ Finish washes out

Where to find
- ☒ Any fabric store
- ☒ Major chain store
- ☒ Stores that carry high quality fabric
- ☒ Fabric club
- ☒ Mail order
- ☒ Wholesale supplier

Brocade

Damask is frequently confused with its cousin, brocade. Both fabrics have complicated floral or figured patterns woven on a jacquard loom. But brocade is heavier than damask and has longer satin floats that snag more easily. Brocade is usually not reversible, while either side of damask may be used as the face. Damask is a single color, while brocade is woven with several colors. Most brocades are made of silk, rayon and/or synthetic fibers, and sometimes dressed up with metallic yarns.

damask

A variety of solid-colored, reversible fabrics woven on a jacquard loom, using a blend of plain and satin weaves to create complex patterns with contrasting luster. Either side of the fabric may be used as the face, but the right side is usually considered to be the side on which the warp floats form the pattern. Damask is flatter and more durable than its cousin, brocade. It may be mercerized or finished using another method to add luster and smoothness. Damask is one of the oldest and most popular fabrics, originally made in China of silk and introduced to the West through Damascus, for which it is named. It is also made of linen, rayon, silk, wool and synthetic fibers.

How to use

Cotton damask has a graceful drape that falls into soft flares. It may be gathered or shirred into a soft fullness. The fabric is easy to cut and sew, but the woven design may require a one-way layout. Use to make table linens, clothing, bedspreads, upholstery and draperies, depending on fabric's weight.

ATTACH SAMPLE HERE

LENGTHWISE GRAIN

The woven design of damask is reversible.

---

double damask
Damask fabric made with an eight-harness satin weave. Fabric has a firm hand. Also called true damask or reversible damask.

single damask
Damask fabric made with a five-harness satin weave.

table damask
A popular cotton tablecloth fabric woven with red or blue warp yarns and white filling yarns, often in a reversible checkerboard pattern. Other versions are bleached white, dyed in pastel shades or woven with colored borders. The fabric is also called two-color damask.

jacquard
Technically, jacquard is not a fabric, but the name of a complex loom used to weave a number of different fabrics with figures and patterns. The term is used rather loosely today to describe any fabric made on a jacquard loom, including damask, heavy brocades and tapestries.

Sewing rating
- ☒ Easy to sew
- ☐ Moderately easy
- ☐ Average
- ☐ Moderately difficult
- ☐ Extremely difficult

Suggested fit
- ☐ Stretch to fit
- ☐ Close-fitting
- ☒ Fitted
- ☒ Semi-fitted
- ☒ Loose-fitting
- ☐ Very loose-fitting

Suggested styles
- ☒ Pleats ☒ Tucks
  - ☒ pressed
  - ☐ unpressed
- ☒ Gathers
  - ☐ limp ☒ soft
  - ☐ full ☐ lofty
  - ☐ bouffant
- ☒ Elasticized shirring
- ☒ Smocked
- ☒ Tailored
- ☐ Shaped with seams to eliminate bulk
- ☐ Lined
- ☐ Unlined
- ☐ Puffed or bouffant
- ☒ Loose and full
- ☒ Soft and flowing
- ☒ Draped
- ☐ Cut on bias
- ☐ Stretch styling

What to expect
- ☐ Difficult to cut out
- ☒ Fabric has one-way
  - ☒ design ☐ luster
  - ☒ weave ☐ nap
- ☒ Fabric is reversible
- ☐ It looks the same on both sides
- ☐ It stretches easily
- ☐ It will not stretch
- ☐ Fabric tears easily
- ☒ It is difficult to tear
- ☐ Fabric will not tear
- ☐ Pins and needles leave holes, marks
- ☐ It is difficult to ease sleeves and curves
- ☐ It tends to pucker
- ☐ It tends to unravel
- ☐ Inner construction shows from outside
- ☐ Machine eats fabric
- ☐ Skipped stitches
- ☐ Layers feed unevenly
- ☐ Multiple layers are difficult to cut, sew
- ☒ It creases easily
- ☐ Won't hold a crease

Cost per yard
- ☐ Less than $5
- ☒ $5 to $10
- ☒ $10 to $15
- ☐ $15 to $20
- ☐ $20 to $25
- ☐ More than $25

Jacquard weave
The jacquard loom uses the three basic weaves — plain, twill and satin — to form complicated floral or figured patterns on a simpler background. The complex loom was developed in France during the early 1800s by Joseph-Marie Jacquard. The loom mechanized figure weaving, an expensive, time-consuming task done by hand. Today's computerized looms allow up to 1,200 yarns to have independent weave action, and are used to make woven and knitted fabrics. It takes a lot of skill and extra time to set up the loom, so it remains an expensive form of weaving.

Wearability
- ☒ Durable ☐ Fragile
- ☐ Strong ☐ Weak
- ☒ It is long-wearing
- ☒ It wears evenly
- ☐ It wears out along seams and folds
- ☐ Seams don't hold up under stress
- ☐ Finish wears off
- ☐ Subject to abrasion
- ☐ It resists abrasion
- ☒ Subject to snags
- ☐ It resists snags
- ☐ Subject to runs
- ☐ It tends to pill
- ☐ It tends to shed
- ☐ It produces lint
- ☐ It attracts lint
- ☐ It attract static
- ☐ It tends to cling
- ☒ It holds its shape
- ☐ It loses its shape
- ☐ It stretches out of shape easily
- ☐ It droops, bags
- ☐ It tends to wrinkle
- ☒ It resists wrinkles
- ☐ It crushes easily
- ☒ Water drops leave spots or marks

Suggested care
- ☐ Dry clean only
- ☐ Do not dry clean
- ☒ Dry clean or wash
- ☐ Gently handwash in lukewarm water
- ☐ Roll in a towel to remove moisture
- ☐ Drip dry
- ☐ Lay flat to dry
- ☒ Machine wash
  - ☐ gentle/delicate
  - ☒ regular/normal
- ☒ Machine dry
  - ☐ cool ☒ normal
  - ☒ perm. press
- ☐ Press damp fabric
- ☒ Press dry fabric
- ☐ Dry iron ☒ Steam
- ☐ Iron on wrong side
- ☐ Use a press cloth
- ☐ Use a needleboard
- ☐ Needs no ironing
- ☐ Do not iron
- ☒ Fabric may shrink
- ☐ May bleed or fade
- ☒ Finish washes out

Where to find
- ☐ Any fabric store
- ☐ Major chain store
- ☒ Stores that carry high quality fabric
- ☒ Fabric club
- ☒ Mail order
- ☒ Wholesale supplier

denim

A densely woven, all-cotton twill fabric made with colored warp yarns and white fillings. The enormously popular fabric is available in many colors, but authentic denim is always dark indigo blue, which fades with age and repeated washings to a pale, muted blue. The strong durable fabric is stiff when new, but it softens with wear and conforms to the body, making it fit like an old shoe. Denim shrinks badly the first time it is washed unless it has been treated to reduce shrinkage. The weight varies from 6- or 8-ounce denim for summer clothing to 16-ounce or more for work clothes.

How to use

Denim has a stiff drape that falls away from the body in wide cones. It works best when shaped with seams to eliminate bulk. Fabric is moderately difficult to cut and sew — flat-felled seams add a challenge and the tight weave may be difficult to ease. Use to make close-fitting, fitted or semi-fitted jeans, skirts, jackets, work clothes and casual wear. Sometimes used for evening wear.

ATTACH SAMPLE HERE

LENGTHWISE GRAIN

Denim has blue warps and white fillings.

Indigo blues

The dark blue warp yarns of the original denim were dyed with indigo, a natural dye extracted from the indigo plant. The blue fabric was difficult to produce, mainly because cotton and indigo are not compatible. Early fabrics had streaks, uneven shading and a tendency to fade and bleed in the wash. Ironically, it was exactly these traits that made denim so popular — the more it faded, the more we liked it. Most denim fabrics are now colored with a synthetic indigo dye that is easier to control.

---

acid-washed denim
Fabric is washed in acid to fade the color.

brushed denim
The fabric is brushed to make it softer.

dungaree
A coarse blue denim used for work clothes.

jean
A twill fabric made of carded cotton yarns, similar to denim but lighter in weight with a finer twill line. Gave its name to blue jeans.

prewashed denim
The fabric is washed to reduce shrinkage and remove excess dye.

stone-washed denim
Denim fabric or garment that has been treated to an abrasive finishing wash, using pumice or pumice-like stones, to produce a worn appearance. The stones soften the fabric as they rub against it. A frosted or iced appearance is produced by coating the stones with bleach.

Sewing rating
- [ ] Easy to sew
- [ ] Moderately easy
- [ ] Average
- [x] Moderately difficult
- [ ] Extremely difficult

Suggested fit
- [ ] Stretch to fit
- [x] Close-fitting
- [x] Fitted
- [x] Semi-fitted
- [ ] Loose-fitting
- [ ] Very loose-fitting

Suggested styles
- [ ] Pleats [x] Tucks
    - [ ] pressed
    - [x] unpressed
- [ ] Gathers
    - [ ] limp [ ] soft
    - [ ] full [ ] lofty
    - [ ] bouffant
- [ ] Elasticized shirring
- [ ] Smocked
- [ ] Tailored
- [x] Shaped with seams to eliminate bulk
- [ ] Lined
- [ ] Unlined
- [ ] Puffed or bouffant
- [ ] Loose and full
- [ ] Soft and flowing
- [ ] Draped
- [ ] Cut on bias
- [ ] Stretch styling

What to expect
- [x] Difficult to cut out
- [ ] Fabric has one-way
    - [ ] design [ ] luster
    - [ ] weave [ ] nap
- [ ] Fabric is reversible
- [ ] It looks the same on both sides
- [ ] It stretches easily
- [x] It will not stretch
- [ ] Fabric tears easily
- [x] It is difficult to tear
- [ ] Fabric will not tear
- [ ] Pins and needles leave holes, marks
- [x] It is difficult to ease sleeves and curves
- [ ] It tends to pucker
- [ ] It tends to unravel
- [ ] Inner construction shows from outside
- [ ] Machine eats fabric
- [ ] Skipped stitches
- [ ] Layers feed unevenly
- [x] Multiple layers are difficult to cut, sew
- [ ] It creases easily
- [x] Won't hold a crease

Cost per yard
- [ ] Less than $5
- [x] $5 to $10
- [ ] $10 to $15
- [ ] $15 to $20
- [ ] $20 to $25
- [ ] More than $25

From workhorse to superstar

Denim gets its name from "serge de Nîmes," a twill fabric first made in Nîmes, France. The rugged fabric literally sailed to America with Christopher Columbus in 1492, as the sails of his ship, the Santa Maria. Denim earned a reputation as a workhorse in the mid-1800s, when Levi Strauss used a modern version of the fabric to make his now-famous blue jeans. With help from Strauss & Co., denim leaped onto the fashion runway in the late 1960s and early 1970s, when baby boomers lived in their worn, faded 501s. It remains in the spotlight.

Wearability
- [x] Durable [ ] Fragile
- [x] Strong [ ] Weak
- [x] It is long-wearing
- [ ] It wears evenly
- [x] It wears out along seams and folds
- [ ] Seams don't hold up under stress
- [ ] Finish wears off
- [x] Subject to abrasion
- [ ] It resists abrasion
- [ ] Subject to snags
- [x] It resists snags
- [ ] Subject to runs
- [ ] It tends to pill
- [ ] It tends to shed
- [ ] It produces lint
- [ ] It attracts lint
- [ ] It attract static
- [ ] It tends to cling
- [x] It holds its shape
- [ ] It loses its shape
- [ ] It stretches out of shape easily
- [ ] It droops, bags
- [ ] It tends to wrinkle
- [x] It resists wrinkles
- [ ] It crushes easily
- [ ] Water drops leave spots or marks

Suggested care
- [ ] Dry clean only
- [ ] Do not dry clean
- [ ] Dry clean or wash
- [ ] Gently handwash in lukewarm water
- [ ] Roll in a towel to remove moisture
- [ ] Drip dry
- [ ] Lay flat to dry
- [x] Machine wash
    - [ ] gentle/delicate
    - [x] regular/normal
- [x] Machine dry
    - [ ] cool [x] normal
    - [ ] perm. press
- [ ] Press damp fabric
- [ ] Press dry fabric
- [ ] Dry iron [ ] Steam
- [ ] Iron on wrong side
- [ ] Use a press cloth
- [ ] Use a needleboard
- [x] Needs no ironing
- [ ] Do not iron
- [x] Fabric may shrink
- [x] May bleed or fade
- [ ] Finish washes out

Where to find
- [x] Any fabric store
- [x] Major chain store
- [x] Stores that carry high quality fabric
- [x] Fabric club
- [x] Mail order
- [x] Wholesale supplier

Flocked dots

Imitation dotted Swiss has flocked dots, made from very short fibers of cotton, wool or rayon, called flock. The small dots and various other shapes are applied to lightweight or sheer fabrics with an adhesive, resin or paste, usually by passing the fabric over brushes or rollers. Adhesives change the hand and drape of the fabric and they may dissolve in the wash, along with the flocks. The dots are subject to abrasion and eventually will rub off. On the plus side, flocked fabrics are inexpensive.

dotted Swiss

A sheer, dainty, plain-weave fabric made of very fine combed cotton and decorated with woven dots at regular intervals. Authentic dotted Swiss is an expensive fabric with hand-knotted dots, made in Switzerland on a swivel or lappet loom. Less costly versions have tufted clipped dots, while cheap imitations have flocked dots. The background fabric is similar to lawn; it may be soft or crisp. It is usually white or pastel with matching or contrasting dots – white, red or pastel dots on a white background are common. The dots are sometimes made of rayon or another fiber to add luster or variety.

How to use

Dotted Swiss has a limp, soft or crisp drape that may be gathered or shirred into delicate, graceful folds or a lofty fullness. Sheer fabrics require special treatment to camouflage seams and facings, adding a degree of difficulty to the simplest of projects. The dots are subject to abrasion. Use to make semi-fitted, loose or very loose dresses, blouses, lingerie, curtains and bridal gowns.

ATTACH SAMPLE HERE

LENGTHWISE GRAIN

Dotted Swiss has hand-tied or woven dots.

clipped dots
Tufted dots made with an extra set of filling yarns inserted into the background fabric at regular intervals. The extra yarns are clipped to produce the dots.

clip-spot lawn
Another name for dotted Swiss.

clipped spots
Fabric woven on a box loom. The extra set of yarns is not woven into the selvages. The term is used interchangeably with clipped dots, but spots are usually larger.

dotted muslin
Another name for dotted Swiss.

swivel dots
Made on a loom with a special attachment that holds tiny shuttles. The shuttles loop extra filling yarns around the warp yarns at regular intervals, then carry the yarns along the fabric's surface to the next spot. The yarns are sheared off between spots. Fabrics are expensive and hard to find.

Sewing rating
- ☐ Easy to sew
- ☐ Moderately easy
- ☒ Average
- ☐ Moderately difficult
- ☐ Extremely difficult

Suggested fit
- ☐ Stretch to fit
- ☐ Close-fitting
- ☒ Fitted
- ☒ Semi-fitted
- ☒ Loose-fitting
- ☐ Very loose-fitting

Suggested styles
- ☐ Pleats ☒ Tucks
  - ☐ pressed
  - ☒ unpressed
- ☒ Gathers
  - ☐ limp ☒ soft
  - ☒ full ☒ lofty
  - ☐ bouffant
- ☒ Elasticized shirring
- ☐ Smocked
- ☐ Tailored
- ☐ Shaped with seams to eliminate bulk
- ☒ Lined
- ☐ Unlined
- ☒ Puffed or bouffant
- ☒ Loose and full
- ☒ Soft and flowing
- ☐ Draped
- ☐ Cut on bias
- ☐ Stretch styling

What to expect
- ☐ Difficult to cut out
- ☐ Fabric has one-way
  - ☐ design ☐ luster
  - ☐ weave ☐ nap
- ☒ Fabric is reversible
- ☒ It looks the same on both sides
- ☐ It stretches easily
- ☐ It will not stretch
- ☐ Fabric tears easily
- ☒ It is difficult to tear
- ☐ Fabric will not tear
- ☒ Pins and needles leave holes, marks
- ☐ It is difficult to ease sleeves and curves
- ☐ It tends to pucker
- ☐ It tends to unravel
- ☒ Inner construction shows from outside
- ☒ Machine eats fabric
- ☐ Skipped stitches
- ☐ Layers feed unevenly
- ☐ Multiple layers are difficult to cut, sew
- ☒ It creases easily
- ☐ Won't hold a crease

Cost per yard
- ☐ Less than $5
- ☐ $5 to $10
- ☐ $10 to $15
- ☒ $15 to $20
- ☒ $20 to $25
- ☒ More than $25

Swiss fabrics

Dotted Swiss is one of many exceptionally fine cotton fabrics made in Switzerland. Swiss mills use the best grades of long-staple, combed cotton to weave sheer organdies, very fine lawns and delicate fabrics like Swiss batiste. Switzerland's reputation for fine fabrics is so strong that the word "Swiss" is used rather loosely today to suggest quality, even on fabrics made somewhere else. Imported Swiss fabrics are usually labeled, in part to justify their higher price. Other well-known fabrics include Swiss cambric, Swiss checks, Swiss embroidery, Swiss mull, Swiss muslin, Swiss pongee and Swiss voile.

Wearability
- ☐ Durable ☒ Fragile
- ☐ Strong ☐ Weak
- ☐ It is long-wearing
- ☐ It wears evenly
- ☐ It wears out along seams and folds
- ☐ Seams don't hold up under stress
- ☐ Finish wears off
- ☒ Subject to abrasion
- ☐ It resists abrasion
- ☒ Subject to snags
- ☐ It resists snags
- ☐ Subject to runs
- ☐ It tends to pill
- ☐ It tends to shed
- ☐ It produces lint
- ☐ It attracts lint
- ☐ It attract static
- ☐ It tends to cling
- ☒ It holds its shape
- ☐ It loses its shape
- ☐ It stretches out of shape easily
- ☐ It droops, bags
- ☒ It tends to wrinkle
- ☐ It resists wrinkles
- ☐ It crushes easily
- ☒ Water drops leave spots or marks

Suggested care
- ☐ Dry clean only
- ☐ Do not dry clean
- ☒ Dry clean or wash
- ☒ Gently handwash in lukewarm water
- ☒ Roll in a towel to remove moisture
- ☐ Drip dry
- ☒ Lay flat to dry
- ☐ Machine wash
  - ☐ gentle/delicate
  - ☐ regular/normal
- ☐ Machine dry
  - ☐ cool ☐ normal
  - ☐ perm. press
- ☒ Press damp fabric
- ☐ Press dry fabric
- ☒ Dry iron ☐ Steam
- ☐ Iron on wrong side
- ☐ Use a press cloth
- ☐ Use a needleboard
- ☐ Needs no ironing
- ☐ Do not iron
- ☒ Fabric may shrink
- ☐ May bleed or fade
- ☒ Finish washes out

Where to find
- ☐ Any fabric store
- ☐ Major chain store
- ☒ Stores that carry high quality fabric
- ☐ Fabric club
- ☒ Mail order
- ☒ Wholesale supplier

double knit

Medium to heavyweight cotton knit, made on a machine with two sets of needles. Both sides of this smooth, flat fabric look the same. Double knit does not stretch in the lengthwise direction and has only a slight, controlled crosswise stretch. It is heavier than cotton interlock and jersey and not as common. Double knit is usually white or solid in color, but it is sometimes printed or embellished with fancy stitches. Some versions have a bit of spandex for added stretch and control. Cotton/polyester blends are easiest to find.

How to use

Cotton double knit falls into moderately soft flares that maintain the shape of the garment. The firm, stable fabric holds its shape well and works best with styles that are shaped with seams to eliminate bulk. Use to make fitted or semi-fitted jackets, dresses, skirts and trousers. Dry clean or machine wash and tumble dry on gentle cycles. Fabric may shrink the first time it is washed, especially in the lengthwise direction.

ATTACH SAMPLE HERE

LENGTHWISE GRAIN

Double knit looks the same on both sides.

Rib knits

Rib knits are made with the rib stitch, which forms lengthwise wales that alternate from side to side. Two rows of needles are used – one knits the wales on the face and the other knits the wales on the back. The ribs can be any width. When every other wale appears on the face, it is called a 1x1 rib or a plain rib. A 2x2 rib, or Swiss rib, has pairs of alternating wales. Rib knits are heavier than plain knits and more durable. They are used for collars and cuffs because of a very elastic crosswise stretch.

---

double cloth
A term used to describe woven fabrics with two distinct layers laced together with an extra set of yarns. The heavy, substantial fabrics are often reversible – the two sides may be made with different weaves, yarns, colors and/or patterns. The layers can't be separated without damaging the fabric.

double-faced knit
A thick, heavy double cloth knitted with different types of stitches on each side, bound together with a separate yarn. The front and back appear different. Some versions are knitted with two colors of yarn, increasing the contrast from one side to the other.

double jersey
Another name for double knit.

flat knit
Any knit that has selvages.

tube knit
A tube-shaped knit that has no selvages.

Sewing rating
- ☐ Easy to sew
- ☐ Moderately easy
- ☒ Average
- ☐ Moderately difficult
- ☐ Extremely difficult

Suggested fit
- ☒ Stretch to fit
- ☒ Close-fitting
- ☒ Fitted
- ☒ Semi-fitted
- ☐ Loose-fitting
- ☐ Very loose-fitting

Suggested styles
- ☐ Pleats ☐ Tucks
  - ☐ pressed
  - ☐ unpressed
- ☐ Gathers
  - ☐ limp ☐ soft
  - ☐ full ☐ lofty
  - ☐ bouffant
- ☒ Elasticized shirring
- ☐ Smocked
- ☐ Tailored
- ☒ Shaped with seams to eliminate bulk
- ☐ Lined
- ☐ Unlined
- ☐ Puffed or bouffant
- ☐ Loose and full
- ☐ Soft and flowing
- ☐ Draped
- ☐ Cut on bias
- ☒ Stretch styling

What to expect
- ☐ Difficult to cut out
- ☐ Fabric has one-way
  - ☐ design ☐ luster
  - ☐ weave ☐ nap
- ☒ Fabric is reversible
- ☒ It looks the same on both sides
- ☒ It stretches easily
- ☐ It will not stretch
- ☐ Fabric tears easily
- ☐ It is difficult to tear
- ☒ Fabric will not tear
- ☐ Pins and needles leave holes, marks
- ☐ It is difficult to ease sleeves and curves
- ☐ It tends to pucker
- ☐ It tends to unravel
- ☐ Inner construction shows from outside
- ☐ Machine eats fabric
- ☐ Skipped stitches
- ☐ Layers feed unevenly
- ☐ Multiple layers are difficult to cut, sew
- ☐ It creases easily
- ☒ Won't hold a crease

Cost per yard
- ☐ Less than $5
- ☒ $5 to $10
- ☒ $10 to $15
- ☒ $15 to $20
- ☐ $20 to $25
- ☐ More than $25

Knit fabrics
Knitting is a method of constructing cloth by interlacing rows of loops together. The smallest unit in knit fabrics is the **stitch**, made by pulling a loop of yarn through another loop. A lengthwise row of loops is called a **wale**. Loops in a wale are formed one under another, using the same needle. A row of crosswise loops is called a **course**. The loops in a course are formed one next to another, using different needles. **Single knits** are made on high-speed knitting machines with one set of needles. **Double knits** are made with two sets of needles, which lace the loops together with a double stitch.

Wearability
- ☒ Durable ☐ Fragile
- ☒ Strong ☐ Weak
- ☒ It is long-wearing
- ☒ It wears evenly
- ☐ It wears out along seams and folds
- ☐ Seams don't hold up under stress
- ☐ Finish wears off
- ☐ Subject to abrasion
- ☐ It resists abrasion
- ☒ Subject to snags
- ☐ It resists snags
- ☐ Subject to runs
- ☒ It tends to pill
- ☐ It tends to shed
- ☐ It produces lint
- ☐ It attracts lint
- ☐ It attract static
- ☐ It tends to cling
- ☒ It holds its shape
- ☐ It loses its shape
- ☐ It stretches out of shape easily
- ☐ It droops, bags
- ☐ It tends to wrinkle
- ☒ It resists wrinkles
- ☐ It crushes easily
- ☐ Water drops leave spots or marks

Suggested care
- ☐ Dry clean only
- ☐ Do not dry clean
- ☒ Dry clean or wash
- ☐ Gently handwash in lukewarm water
- ☐ Roll in a towel to remove moisture
- ☐ Drip dry
- ☒ Lay flat to dry
- ☒ Machine wash
  - ☒ gentle/delicate
  - ☐ regular/normal
- ☒ Machine dry
  - ☐ cool ☒ normal
  - ☐ perm. press
- ☐ Press damp fabric
- ☐ Press dry fabric
- ☐ Dry iron ☐ Steam
- ☐ Iron on wrong side
- ☐ Use a press cloth
- ☐ Use a needleboard
- ☒ Needs no ironing
- ☐ Do not iron
- ☒ Fabric may shrink
- ☐ May bleed or fade
- ☐ Finish washes out

Where to find
- ☐ Any fabric store
- ☒ Major chain store
- ☒ Stores that carry high quality fabric
- ☐ Fabric club
- ☒ Mail order
- ☒ Wholesale supplier

Warp-faced twill

Cotton drill is an example of a warp-faced twill, an uneven twill that has more warp yarns than filling yarns on the face. Because warp yarns are stronger than filling yarns, a warp-faced twill will wear better than an even twill, which has a balanced number of warp yarns on both sides. It's easy to tell the difference. An even twill has a diagonal twill line on both sides, while a warp-faced twill has a diagonal line on the face and a plain back. A 3x1 twill is even tougher than a 2x1 twill.

drill

Durable, tightly woven cotton twill, usually made with coarse carded yarns. The firm, strong fabric is medium to heavy in weight with a stiff, rough hand. It often looks like unbleached, undyed denim, but the fabric may also be bleached, dyed or printed. Drill gets its name from the Greek word *drillich*, meaning three warp threads. It is made with a left-hand, warp-faced twill weave that repeats on three threads, usually with a 2x1 (two-up, one-down) pattern. Other versions are made with a 3x1 twill or with a herringbone or shadow stripe effect.

How to use

Drill has a stiff drape that falls away from the body in wide cones. It works best with styles that are shaped with seams to eliminate bulk. Drill is easy to cut and sew, but the tight weave may be difficult to ease. Lighter weights are used to make trouser pockets and shoe linings. Use heavier weights to make close-fitting, fitted or semi-fitted trousers, uniforms and other work clothes. Machine wash and tumble dry, but beware of shrinkage. Needs no ironing.

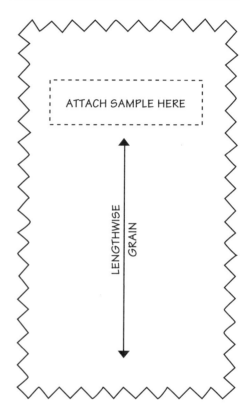

Drill looks like unbleached, undyed denim.

---

even twill
Any twill fabric that has the same number of warp yarns above and below the filling yarns. For example, a 2x2 twill (two-up, two-down) has two warp yarns that pass over one filling yarn; the next two warp yarns pass under the same filling yarn. It is also called a balanced twill.

filling-faced twill
Uncommon twill that has more filling yarns than warp yarns on the fabric's face.

left-hand twill
Any twill with a diagonal line on the face that runs from the lower right to the upper left. Not as common as right-hand twill.

three-harness twill
The simplest twill, which repeats on three threads in a 2x1 (two-up, one-down) weave.

right-hand twill
Any twill with a diagonal line on the face that runs from the lower left to the upper right. Most twills are made this way.

Sewing rating
- ☐ Easy to sew
- ☒ Moderately easy
- ☐ Average
- ☐ Moderately difficult
- ☐ Extremely difficult

Suggested fit
- ☐ Stretch to fit
- ☒ Close-fitting
- ☒ Fitted
- ☒ Semi-fitted
- ☐ Loose-fitting
- ☐ Very loose-fitting

Suggested styles
- ☐ Pleats ☒ Tucks
  - ☐ pressed
  - ☒ unpressed
- ☐ Gathers
  - ☐ limp ☐ soft
  - ☐ full ☐ lofty
  - ☐ bouffant
- ☐ Elasticized shirring
- ☐ Smocked
- ☐ Tailored
- ☒ Shaped with seams to eliminate bulk
- ☐ Lined
- ☐ Unlined
- ☐ Puffed or bouffant
- ☐ Loose and full
- ☐ Soft and flowing
- ☐ Draped
- ☐ Cut on bias
- ☐ Stretch styling

What to expect
- ☒ Difficult to cut out
- ☒ Fabric has one-way
  - ☐ design ☐ luster
  - ☒ weave ☐ nap
- ☐ Fabric is reversible
- ☐ It looks the same on both sides
- ☐ It stretches easily
- ☒ It will not stretch
- ☐ Fabric tears easily
- ☒ It is difficult to tear
- ☒ Fabric will not tear
- ☐ Pins and needles leave holes, marks
- ☒ It is difficult to ease sleeves and curves
- ☐ It tends to pucker
- ☐ It tends to unravel
- ☐ Inner construction shows from outside
- ☐ Machine eats fabric
- ☐ Skipped stitches
- ☐ Layers feed unevenly
- ☒ Multiple layers are difficult to cut, sew
- ☐ It creases easily
- ☒ Won't hold a crease

Cost per yard
- ☒ Less than $5
- ☒ $5 to $10
- ☐ $10 to $15
- ☐ $15 to $20
- ☐ $20 to $25
- ☐ More than $25

The twill weave
Drill is a good example of twill, the strongest of the basic weaves. Most twill fabrics have a woven diagonal rib, called a twill line, on one or both sides. A typical 2x1 twill weave is made with a two-up, one-down pattern, meaning the warp (lengthwise) yarns travel over two filling (crosswise) yarns, under one and over two more. On each row, the filling yarn shifts over one warp yarn to form the diagonal line. The pattern can be modified to create steep, reclining, broken or zigzag lines. Twill is a firm, durable weave, but fabrics tend to shrink, fray and wear out along edges and folds.

Wearability
- ☒ Durable ☐ Fragile
- ☒ Strong ☐ Weak
- ☒ It is long-wearing
- ☐ It wears evenly
- ☒ It wears out along seams and folds
- ☐ Seams don't hold up under stress
- ☐ Finish wears off
- ☐ Subject to abrasion
- ☒ It resists abrasion
- ☐ Subject to snags
- ☒ It resists snags
- ☐ Subject to runs
- ☐ It tends to pill
- ☐ It tends to shed
- ☐ It produces lint
- ☐ It attracts lint
- ☐ It attract static
- ☐ It tends to cling
- ☒ It holds its shape
- ☐ It loses its shape
- ☐ It stretches out of shape easily
- ☐ It droops, bags
- ☐ It tends to wrinkle
- ☒ It resists wrinkles
- ☐ It crushes easily
- ☐ Water drops leave spots or marks

Suggested care
- ☐ Dry clean only
- ☐ Do not dry clean
- ☐ Dry clean or wash
- ☐ Gently handwash in lukewarm water
- ☐ Roll in a towel to remove moisture
- ☒ Drip dry
- ☐ Lay flat to dry
- ☒ Machine wash
  - ☐ gentle/delicate
  - ☒ regular/normal
- ☒ Machine dry
  - ☐ cool ☒ normal
  - ☐ perm. press
- ☐ Press damp fabric
- ☐ Press dry fabric
- ☐ Dry iron ☐ Steam
- ☐ Iron on wrong side
- ☐ Use a press cloth
- ☐ Use a needleboard
- ☒ Needs no ironing
- ☐ Do not iron
- ☒ Fabric may shrink
- ☐ May bleed or fade
- ☒ Finish washes out

Where to find
- ☐ Any fabric store
- ☒ Major chain store
- ☐ Stores that carry high quality fabric
- ☐ Fabric club
- ☒ Mail order
- ☒ Wholesale supplier

duck

A large group of coarse, rough, plain-weave cotton fabrics used for industrial and utility purposes, such as conveyor belts and tote bags. The term is used interchangeably with canvas, but duck usually refers to a heavier, coarser fabric while canvas tends to be smoother and more tightly woven. The strong, durable fabric is sometimes treated to resist mildew and repel water and used to make tents, awnings and similar outdoor items. Duck comes in a variety of widths and weights. It is often labeled by ounces per yard, such as 12-ounce duck.

Duck is coarser and heavier than canvas.

How to use

Duck has a stiff, heavy drape that falls into wide cones. The macho fabric is deceptively difficult to handle: Dull or wimpy shears make cutting a struggle and multiple layers may be difficult to feed under the machine's presser foot, a problem if you want to make flat-felled seams. Don't be surprised if the fabric proves to be stronger than your toughest needle. Use to make luggage and tote bags, awnings, tents and seats/backs of outdoor furniture.

Birds of a feather

The term "duck" can be traced to the mid-19th century, when canvas was made of flax (linen) and imported into the United States, mostly from England and Scotland. The fabrics were available in several weights, so a labeling system was devised to keep them straight. Lighter weights bore the stencil mark of a raven, while heavier fabrics pictured a duck. "Duck" became slang for heavy canvas. When a heavy canvas was first made in the United States, the name stuck, and it was called duck.

---

army duck
Duck fabric made with a plain weave using medium to heavy ply yarns. Similar to numbered duck, but lighter in weight.

biscuit duck
A numbered duck used by commercial bakeries during the preparation of dough.

flat duck
A cotton duck that is light to medium in weight, made with single warp yarns woven in pairs and single or ply filling yarns. The fabric has many more warp yarns per inch than filling yarns. Fabric's weight ranges from 6 to 15 ounces per yard, based on a width of 29 inches. Also called ounce duck.

numbered duck
Closely woven, strong, durable fabrics made with a plain weave and medium to heavy ply yarns. Fabrics are numbered from 1 to 12, according to weight. The actual weight is figured by subtracting the number from 19. For example, a yard of No. 1 duck weighs 18 ounces, based on a width of 22 inches.

Sewing rating
- [ ] Easy to sew
- [ ] Moderately easy
- [ ] Average
- [x] Moderately difficult
- [ ] Extremely difficult

Suggested fit
- [ ] Stretch to fit
- [x] Close-fitting
- [x] Fitted
- [x] Semi-fitted
- [ ] Loose-fitting
- [ ] Very loose-fitting

Suggested styles
- [ ] Pleats [ ] Tucks
  - [ ] pressed
  - [ ] unpressed
- [ ] Gathers
  - [ ] limp [ ] soft
  - [ ] full [ ] lofty
  - [ ] bouffant
- [ ] Elasticized shirring
- [ ] Smocked
- [ ] Tailored
- [x] Shaped with seams to eliminate bulk
- [ ] Lined
- [ ] Unlined
- [ ] Puffed or bouffant
- [ ] Loose and full
- [ ] Soft and flowing
- [ ] Draped
- [ ] Cut on bias
- [ ] Stretch styling

What to expect
- [x] Difficult to cut out
- [ ] Fabric has one-way
  - [ ] design [ ] luster
  - [ ] weave [ ] nap
- [x] Fabric is reversible
- [x] It looks the same on both sides
- [ ] It stretches easily
- [x] It will not stretch
- [ ] Fabric tears easily
- [x] It is difficult to tear
- [ ] Fabric will not tear
- [ ] Pins and needles leave holes, marks
- [x] It is difficult to ease sleeves and curves
- [ ] It tends to pucker
- [ ] It tends to unravel
- [ ] Inner construction shows from outside
- [ ] Machine eats fabric
- [x] Skipped stitches
- [x] Layers feed unevenly
- [x] Multiple layers are difficult to cut, sew
- [ ] It creases easily
- [x] Won't hold a crease

Cost per yard
- [x] Less than $5
- [x] $5 to $10
- [x] $10 to $15
- [ ] $15 to $20
- [ ] $20 to $25
- [ ] More than $25

Sun damage
Duck and canvas fabrics are often used to make awnings, tents and the seats of deck chairs and other outdoor furniture. But the rugged fabrics may not be as durable as expected. Cotton grows weaker and eventually breaks down when exposed to extended periods of sunlight. After only two weeks in the hot sun, the fiber may lose as much as 50 percent of its strength. With prolonged exposure to the sun's ultraviolet rays, cotton turns yellow and grows even weaker. Sun damage is exaggerated when moisture is present and by some dyes. When properly stored, cotton will retain most of its strength and appearance.

Wearability
- [x] Durable [ ] Fragile
- [x] Strong [ ] Weak
- [x] It is long-wearing
- [x] It wears evenly
- [ ] It wears out along seams and folds
- [ ] Seams don't hold up under stress
- [x] Finish wears off
- [ ] Subject to abrasion
- [x] It resists abrasion
- [ ] Subject to snags
- [x] It resists snags
- [ ] Subject to runs
- [ ] It tends to pill
- [ ] It tends to shed
- [ ] It produces lint
- [ ] It attracts lint
- [ ] It attract static
- [ ] It tends to cling
- [x] It holds its shape
- [ ] It loses its shape
- [ ] It stretches out of shape easily
- [ ] It droops, bags
- [ ] It tends to wrinkle
- [x] It resists wrinkles
- [ ] It crushes easily
- [ ] Water drops leave spots or marks

Suggested care
- [ ] Dry clean only
- [ ] Do not dry clean
- [ ] Dry clean or wash
- [ ] Gently handwash in lukewarm water
- [ ] Roll in a towel to remove moisture
- [x] Drip dry
- [ ] Lay flat to dry
- [x] Machine wash
  - [ ] gentle/delicate
  - [x] regular/normal
- [x] Machine dry
  - [ ] cool [x] normal
  - [ ] perm. press
- [ ] Press damp fabric
- [ ] Press dry fabric
- [ ] Dry iron [ ] Steam
- [ ] Iron on wrong side
- [ ] Use a press cloth
- [ ] Use a needleboard
- [x] Needs no ironing
- [ ] Do not iron
- [x] Fabric may shrink
- [ ] May bleed or fade
- [x] Finish washes out

Where to find
- [x] Any fabric store
- [x] Major chain store
- [ ] Stores that carry high quality fabric
- [ ] Fabric club
- [x] Mail order
- [x] Wholesale supplier

Lace

Eyelet is sometimes confused with lace, especially when it is made in narrow widths and used for trim. But lace is different — unlike eyelet and embroidered cloths, lace is made without the aid of a background fabric. Lace fabrics and trims are formed by twisting together or knotting a network of threads to produce delicate openwork designs. Lace originally was made by hand, but today it is usually made by machine. Handmade lace has an irregular mesh and the pattern repeats are not exactly alike.

eyelet

Any fabric that has been decorated with an openwork design, most often in the form of small holes surrounded by embroidery-like stitching produced on a Schiffli embroidery machine. The background is usually a firm, lightweight, smooth, plain fabric, such as batiste, broadcloth, lawn or organdy. The fabric is usually mercerized, but it may have a soft or crisp hand, depending on any other finishes. Eyelet is most often white, cream or pastel in color, with a matching design, but it is sometimes embellished with a contrasting color. Cotton/polyester fabrics are more common and less expensive than all-cotton versions.

How to use

Eyelet has a soft to crisp drape that falls into moderately soft flares. It may be gathered into a soft or lofty fullness. The fabric may be delicate or quite durable, but the decorative stitching tends to come undone. Use to make semi-fitted or loose-fitting summer blouses, dresses, wedding gowns, baby clothes, children's dresses and curtains. Machine wash and tumble dry.

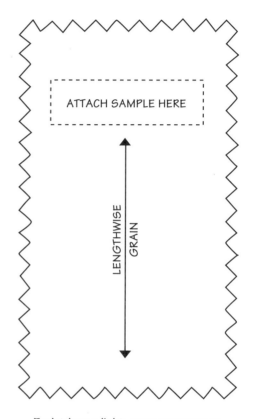

Eyelet has a light, open appearance.

---

embroidery
Any embellishment or decoration of thread, yarn or other flexible material on a fabric or leather background. Embroidery requires a background fabric, while lace usually does not. Embroidery was originally done by hand, but most of today's embroidered fabrics are produced by machine.

needlepoint
Any embroidery work done on a background of heavy scrim, mesh or canvas.

needlework
A general term for work performed with a needle, including embroidery, knitting, some lace work and hand sewing.

Schiffli embroidery machine
A machine that uses 684 or 1,026 needles and multiple threads to embroider fabrics. A perforated roll guides the placement of each stitch. The embroidery can be done on any fabric, but the background is usually a sheer fabric, such as organdy. Net backing is used to produce lace-like fabrics.

Sewing rating
- ☐ Easy to sew
- ☒ Moderately easy
- ☐ Average
- ☐ Moderately difficult
- ☐ Extremely difficult

Suggested fit
- ☐ Stretch to fit
- ☒ Close-fitting
- ☒ Fitted
- ☒ Semi-fitted
- ☒ Loose-fitting
- ☐ Very loose-fitting

Suggested styles
- ☒ Pleats ☒ Tucks
  - ☒ pressed
  - ☒ unpressed
- ☒ Gathers
  - ☒ limp ☒ soft
  - ☒ full ☐ lofty
  - ☐ bouffant
- ☒ Elasticized shirring
- ☐ Smocked
- ☐ Tailored
- ☐ Shaped with seams to eliminate bulk
- ☒ Lined
- ☐ Unlined
- ☐ Puffed or bouffant
- ☒ Loose and full
- ☒ Soft and flowing
- ☒ Draped
- ☐ Cut on bias
- ☐ Stretch styling

What to expect
- ☐ Difficult to cut out
- ☒ Fabric has one-way
  - ☒ design ☐ luster
  - ☐ weave ☐ nap
- ☐ Fabric is reversible
- ☐ It looks the same on both sides
- ☐ It stretches easily
- ☐ It will not stretch
- ☐ Fabric tears easily
- ☒ It is difficult to tear
- ☐ Fabric will not tear
- ☐ Pins and needles leave holes, marks
- ☐ It is difficult to ease sleeves and curves
- ☐ It tends to pucker
- ☐ It tends to unravel
- ☒ Inner construction shows from outside
- ☐ Machine eats fabric
- ☐ Skipped stitches
- ☐ Layers feed unevenly
- ☐ Multiple layers are difficult to cut, sew
- ☒ It creases easily
- ☐ Won't hold a crease

Cost per yard
- ☐ Less than $5
- ☒ $5 to $10
- ☒ $10 to $15
- ☐ $15 to $20
- ☐ $20 to $25
- ☐ More than $25

The lappet weave
Embellished fabrics are sometimes made with a lappet weave, which uses a special needle attached to the loom to produce a figured or embroidered design on a smooth, plain-weave background fabric. The needle attachment carries its own supply of warp yarns from one side of the loom to the other, stopping at regular intervals to embroider a design into the fabric. The design is fastened or knotted to the background fabric and loose floats on the back of the fabric are cut. The ornamental design is usually quite durable, but fabrics are expensive and not very common.

Wearability
- ☐ Durable ☐ Fragile
- ☐ Strong ☐ Weak
- ☐ It is long-wearing
- ☐ It wears evenly
- ☐ It wears out along seams and folds
- ☐ Seams don't hold up under stress
- ☐ Finish wears off
- ☒ Subject to abrasion
- ☐ It resists abrasion
- ☒ Subject to snags
- ☐ It resists snags
- ☐ Subject to runs
- ☐ It tends to pill
- ☐ It tends to shed
- ☐ It produces lint
- ☐ It attracts lint
- ☐ It attract static
- ☐ It tends to cling
- ☒ It holds its shape
- ☐ It loses its shape
- ☐ It stretches out of shape easily
- ☐ It droops, bags
- ☒ It tends to wrinkle
- ☐ It resists wrinkles
- ☐ It crushes easily
- ☐ Water drops leave spots or marks

Suggested care
- ☐ Dry clean only
- ☐ Do not dry clean
- ☐ Dry clean or wash
- ☒ Gently handwash in lukewarm water
- ☒ Roll in a towel to remove moisture
- ☒ Drip dry
- ☐ Lay flat to dry
- ☒ Machine wash
  - ☒ gentle/delicate
  - ☐ regular/normal
- ☒ Machine dry
  - ☒ cool ☐ normal
  - ☐ perm. press
- ☐ Press damp fabric
- ☒ Press dry fabric
- ☐ Dry iron ☒ Steam
- ☐ Iron on wrong side
- ☐ Use a press cloth
- ☐ Use a needleboard
- ☐ Needs no ironing
- ☐ Do not iron
- ☒ Fabric may shrink
- ☐ May bleed or fade
- ☒ Finish washes out

Where to find
- ☒ Any fabric store
- ☒ Major chain store
- ☒ Stores that carry high quality fabric
- ☐ Fabric club
- ☐ Mail order
- ☒ Wholesale supplier

flannel

Downy soft, warm fabric that is brushed on one or both sides to raise the nap. Flannel is usually cotton or a cotton/synthetic blend, woven with a twill or plain weave. The weight varies from lightweight pajama fabrics to diaper cloths to heavy shirtings that feel almost like felt. Flannel is considered to be a winter fabric because the brushing process adds loft to the fibers, creating air space that traps heat. The popular fabric may be dyed, bleached or printed. Heavy flannels often have plaid designs that may be woven or printed. Some fabrics are mercerized and/or treated to control shrinkage.

How to use

Flannel has a graceful drape that falls close to the body in soft flares. It may be gathered or shirred into a moderate fullness. Use lightweight flannels for baby clothes, shirts, dresses and sleepwear. Use heavier flannels for shirts, linings and lightweight casual jackets. Flannel soils easily and the napped surface eventually wears off. Machine wash and dry, but beware of shrinkage.

Brushed cotton flannel is soft and warm.

Brushed cotton

Flannel's soft, fuzzy surface is produced by feeding the fabric over a revolving brush to raise the fiber ends to the surface, creating a slight fuzz or a thick nap that may be very compact or loosely formed. The fibers are trimmed, brushed flat or left in a raised, upright position. The finished fabric is warm because brushing creates air space that traps heat. Brushing also may be used to conceal loose weaves and inferior fabrics. All brushed fabrics tend to shed and the nap eventually flattens.

Canton flannel
A soft twill fabric with a long raised nap on the back, made of carded cotton and soft, heavy filling yarns. May be bleached, dyed or printed. Used for sleepwear, underwear, interlinings and sportswear. Glove flannel is a heavily napped variation with colored stripes, used primarily for work gloves.

diaper flannel
A plain cotton fabric brushed on both sides to increase softness and absorbency. It is usually made in 27- or 30-inch widths.

flannelette
A lightweight, plain-weave cotton flannel with a light nap on one side.

outing flannel
A general term for twilled or plain-weave cotton fabric napped on one or both sides. May be white, a solid color, printed or have woven stripes, checks or plaids. Available in various weights and qualities and used for for shirts, sleepwear, diapers, baby clothes and interlinings. Formerly called domet.

Sewing rating
- ☒ Easy to sew
- ☐ Moderately easy
- ☐ Average
- ☐ Moderately difficult
- ☐ Extremely difficult

Suggested fit
- ☐ Stretch to fit
- ☐ Close-fitting
- ☒ Fitted
- ☒ Semi-fitted
- ☒ Loose-fitting
- ☐ Very loose-fitting

Suggested styles
- ☐ Pleats ☒ Tucks
  - ☐ pressed
  - ☒ unpressed
- ☒ Gathers
  - ☒ limp ☒ soft
  - ☒ full ☐ lofty
  - ☐ bouffant
- ☒ Elasticized shirring
- ☐ Smocked
- ☐ Tailored
- ☐ Shaped with seams to eliminate bulk
- ☐ Lined
- ☐ Unlined
- ☐ Puffed or bouffant
- ☒ Loose and full
- ☒ Soft and flowing
- ☐ Draped
- ☐ Cut on bias
- ☐ Stretch styling

What to expect
- ☐ Difficult to cut out
- ☒ Fabric has one-way
  - ☐ design ☐ luster
  - ☐ weave ☒ nap
- ☐ Fabric is reversible
- ☐ It looks the same on both sides
- ☐ It stretches easily
- ☐ It will not stretch
- ☒ Fabric tears easily
- ☐ It is difficult to tear
- ☐ Fabric will not tear
- ☐ Pins and needles leave holes, marks
- ☐ It is difficult to ease sleeves and curves
- ☐ It tends to pucker
- ☐ It tends to unravel
- ☐ Inner construction shows from outside
- ☐ Machine eats fabric
- ☐ Skipped stitches
- ☐ Layers feed unevenly
- ☐ Multiple layers are difficult to cut, sew
- ☐ It creases easily
- ☒ Won't hold a crease

Cost per yard
- ☒ Less than $5
- ☒ $5 to $10
- ☐ $10 to $15
- ☐ $15 to $20
- ☐ $20 to $25
- ☐ More than $25

Flame-retardant finishes
Some cotton fabrics, including most flannels, carry a label that says "Not intended for use as children's sleepwear." This means the fabric failed the industry's burn test, established in response to the Flammable Fabrics Act of 1953, which led to mandatory standards for certain textile items, including upholstery, carpets and children's sleepwear. To comply with the law, some fabrics are given a flame-retardant finish, which slows the rate of ignition and flames, but does not prevent burning. The finish must be nontoxic, noncarcinogenic and hold up for at least 50 washings. Treated fabrics are more costly.

Wearability
- ☒ Durable ☐ Fragile
- ☐ Strong ☐ Weak
- ☐ It is long-wearing
- ☐ It wears evenly
- ☒ It wears out along seams and folds
- ☐ Seams don't hold up under stress
- ☒ Finish wears off
- ☒ Subject to abrasion
- ☐ It resists abrasion
- ☐ Subject to snags
- ☒ It resists snags
- ☐ Subject to runs
- ☐ It tends to pill
- ☒ It tends to shed
- ☒ It produces lint
- ☐ It attracts lint
- ☐ It attract static
- ☐ It tends to cling
- ☒ It holds its shape
- ☐ It loses its shape
- ☐ It stretches out of shape easily
- ☐ It droops, bags
- ☐ It tends to wrinkle
- ☒ It resists wrinkles
- ☐ It crushes easily
- ☐ Water drops leave spots or marks

Suggested care
- ☐ Dry clean only
- ☐ Do not dry clean
- ☐ Dry clean or wash
- ☐ Gently handwash in lukewarm water
- ☐ Roll in a towel to remove moisture
- ☐ Drip dry
- ☐ Lay flat to dry
- ☒ Machine wash
  - ☐ gentle/delicate
  - ☒ regular/normal
- ☒ Machine dry
  - ☐ cool ☒ normal
  - ☐ perm. press
- ☐ Press damp fabric
- ☒ Press dry fabric
- ☒ Dry iron ☐ Steam
- ☒ Iron on wrong side
- ☐ Use a press cloth
- ☐ Use a needleboard
- ☐ Needs no ironing
- ☐ Do not iron
- ☒ Fabric may shrink
- ☐ May bleed or fade
- ☒ Finish washes out

Where to find
- ☒ Any fabric store
- ☒ Major chain store
- ☒ Stores that carry high quality fabric
- ☒ Fabric club
- ☒ Mail order
- ☒ Wholesale supplier

Cheesecloth

Cheesecloth is often mistaken for gauze, but the two fabrics are different. Cheesecloth is a soft, open-weave fabric made of carded cotton yarns and a plain weave. It is not as stable as authentic gauze, which is made with a variation of the leno weave. Cheesecloth originally was used to wrap pressed cheese, butter and meat, or to strain milk and jelly. It is now used for dust cloths, surgical dressings and book bindings. A one-yard width of cheesecloth is called tobacco cloth.

gauze

A limp, soft cotton fabric made with a loose open weave. Authentic gauze is made with a variation of the leno weave, but the plain weave may also be used. The lightweight, thin fabric is usually sheer, but heavier versions may be more opaque. It is most often made with coarse carded yarns, but fine combed yarns may be used to make more costly versions. Fancy gauze fabrics are made of silk, rayon and wool. Gauze is named for the city of Gaza in the Near East, where it was first made thousands of years ago. In France, gauze is known as gaze. The term is sometimes used to describe a very sheer knitted fabric.

How to use

Gauze has a soft, limp drape that falls close to the body. It may be gathered into a soft fullness, but the fabric won't hold pleats. Gauze ravels easily and tends to snag. It is also subject to yarn slippage and has a pronounced bias stretch. Use to make loose-fitting or very loose-fitting dresses, blouses and casual clothes and curtains. Gauze doesn't wrinkle much, but it shrinks badly.

ATTACH SAMPLE HERE

LENGTHWISE GRAIN

Gauze is made with a loose open weave.

---

leno
A loosely woven, lightweight cotton fabric made with the leno weave, which makes the fabric stronger and less prone to slippage and distortion. The fabric is sometimes incorrectly called gauze. Leno is airy and open, making it a good choice for summer blouses, casual garments and curtains.

marquisette
A crisp, sheer, lightweight fabric with a square, open mesh, made with the leno weave of cotton or synthetic fibers. It was once used for mosquito netting, but today it is more often used to make underlinings, interfacings, curtains and trimmings. Silk versions are used for evening wear.

scrim
An open-mesh, plain-weave cotton fabric made from carded or combed yarns. Scrim is usually bleached and sized to add body.

surgical gauze
Soft, narrow cotton gauze made with the plain weave and used for dressing wounds.

Sewing rating
- [ ] Easy to sew
- [ ] Moderately easy
- [x] Average
- [ ] Moderately difficult
- [ ] Extremely difficult

Suggested fit
- [ ] Stretch to fit
- [ ] Close-fitting
- [ ] Fitted
- [x] Semi-fitted
- [x] Loose-fitting
- [x] Very loose-fitting

Suggested styles
- [ ] Pleats [ ] Tucks
  - [ ] pressed
  - [ ] unpressed
- [x] Gathers
  - [x] limp [x] soft
  - [ ] full [ ] lofty
  - [ ] bouffant
- [x] Elasticized shirring
- [ ] Smocked
- [ ] Tailored
- [ ] Shaped with seams to eliminate bulk
- [ ] Lined
- [ ] Unlined
- [ ] Puffed or bouffant
- [x] Loose and full
- [x] Soft and flowing
- [x] Draped
- [ ] Cut on bias
- [ ] Stretch styling

What to expect
- [ ] Difficult to cut out
- [ ] Fabric has one-way
  - [ ] design [ ] luster
  - [ ] weave [ ] nap
- [x] Fabric is reversible
- [x] It looks the same on both sides
- [x] It stretches easily
- [ ] It will not stretch
- [x] Fabric tears easily
- [ ] It is difficult to tear
- [ ] Fabric will not tear
- [ ] Pins and needles leave holes, marks
- [ ] It is difficult to ease sleeves and curves
- [ ] It tends to pucker
- [ ] It tends to unravel
- [x] Inner construction shows from outside
- [x] Machine eats fabric
- [ ] Skipped stitches
- [ ] Layers feed unevenly
- [ ] Multiple layers are difficult to cut, sew
- [ ] It creases easily
- [x] Won't hold a crease

Cost per yard
- [x] Less than $5
- [x] $5 to $10
- [ ] $10 to $15
- [ ] $15 to $20
- [ ] $20 to $25
- [ ] More than $25

Leno weave

The leno weave adds stability to loose open weaves. It is made with paired warp yarns — one warp is positioned like the warp of a plain weave, while the other warp yarn passes over and under the filling yarns from the other side. The gauze weave is a variation of the leno weave, made by twisting the paired warp yarns around the filling yarns in a figure-8 arrangement. This prevents yarn slippage and adds strength and stability to the fabric. The leno weave is most often used to make drapery fabrics and thermal blankets. It is sometimes called a doup weave.

Wearability
- [ ] Durable [x] Fragile
- [ ] Strong [ ] Weak
- [ ] It is long-wearing
- [x] It wears evenly
- [ ] It wears out along seams and folds
- [x] Seams don't hold up under stress
- [ ] Finish wears off
- [x] Subject to abrasion
- [ ] It resists abrasion
- [x] Subject to snags
- [ ] It resists snags
- [ ] Subject to runs
- [ ] It tends to pill
- [ ] It tends to shed
- [ ] It produces lint
- [ ] It attracts lint
- [ ] It attract static
- [ ] It tends to cling
- [ ] It holds its shape
- [x] It loses its shape
- [x] It stretches out of shape easily
- [x] It droops, bags
- [ ] It tends to wrinkle
- [x] It resists wrinkles
- [ ] It crushes easily
- [ ] Water drops leave spots or marks

Suggested care
- [ ] Dry clean only
- [ ] Do not dry clean
- [ ] Dry clean or wash
- [ ] Gently handwash in lukewarm water
- [ ] Roll in a towel to remove moisture
- [ ] Drip dry
- [ ] Lay flat to dry
- [x] Machine wash
  - [x] gentle/delicate
  - [ ] regular/normal
- [x] Machine dry
  - [x] cool [ ] normal
  - [ ] perm. press
- [ ] Press damp fabric
- [x] Press dry fabric
- [ ] Dry iron [x] Steam
- [ ] Iron on wrong side
- [ ] Use a press cloth
- [ ] Use a needleboard
- [x] Needs no ironing
- [ ] Do not iron
- [x] Fabric may shrink
- [ ] May bleed or fade
- [x] Finish washes out

Where to find
- [x] Any fabric store
- [x] Major chain store
- [ ] Stores that carry high quality fabric
- [ ] Fabric club
- [x] Mail order
- [x] Wholesale supplier

gingham

A lightweight fabric woven with alternating white and colored yarns to form checks or stripes. This plain-weave fabric is usually made from cotton or a cotton/polyester blend. Fabrics vary in weight, quality and price. Versions made with combed single yarns are softer, finer and more costly than versions made with carded yarns. The fabric is usually lightly sized. When the check pattern is printed, it is called imitation gingham. There are conflicting accounts of the fabric's origin, but it is most likely named for Guingamp, a town in Brittany, France, where fabrics of this type were once made.

Yarn-dyed fabric

Yarn-dyed fabric is made with yarns that are dyed before they are woven into cloth — a method used to produce plaids, stripes, checks, tweeds and irridescent fabrics. Gingham's colored checks are formed where dyed yarns intersect. Lighter colored checks are formed where the dyed yarns cross white yarns and white checks are formed where the white yarns intersect. Yarn-dyed fabrics usually are reversible and they tend to be more colorfast because the dye penetrates to the yarn's core.

How to use

Gingham has a soft or crisp drape that falls close to the body in moderately soft flares. The fabric is easy to sew and may be pleated, gathered, shirred or smocked. Inexpensive versions are used to make dresses, pajamas, aprons and close-fitting sample garments, while better qualities are used for suits, dresses, pajamas and kitchen curtains. Machine wash and tumble dry.

ATTACH SAMPLE HERE

LENGTHWISE GRAIN

Gingham has dyed and white yarns.

---

carded gingham
Lower grade fabric made of carded yarns.

chambray gingham
Fine fabric with a lustrous finish that is often piece-dyed, rather than yarn-dyed.

combed gingham
High quality fabric made of combed yarns.

madras gingham
Made with fine yarns and fancy weaves.

nurses' gingham
Has woven stripes instead of checks.

Scotch gingham
The very finest gingham, made in Scotland.

tissue gingham
Very soft, lightweight gingham made with fine, combed single yarns. It often has stripes instead of checks.

zephyr gingham
Very fine, soft gingham, made with combed single yarns. Heavier than tissue gingham.

Sewing rating
- ☒ Easy to sew
- ☐ Moderately easy
- ☐ Average
- ☐ Moderately difficult
- ☐ Extremely difficult

Suggested fit
- ☐ Stretch to fit
- ☒ Close-fitting
- ☒ Fitted
- ☒ Semi-fitted
- ☒ Loose-fitting
- ☐ Very loose-fitting

Suggested styles
- ☒ Pleats ☒ Tucks
  - ☒ pressed
  - ☐ unpressed
- ☒ Gathers
  - ☐ limp ☒ soft
  - ☒ full ☐ lofty
  - ☐ bouffant
- ☒ Elasticized shirring
- ☒ Smocked
- ☐ Tailored
- ☐ Shaped with seams to eliminate bulk
- ☐ Lined
- ☐ Unlined
- ☐ Puffed or bouffant
- ☒ Loose and full
- ☒ Soft and flowing
- ☒ Draped
- ☒ Cut on bias
- ☐ Stretch styling

What to expect
- ☐ Difficult to cut out
- ☐ Fabric has one-way
  - ☐ design ☐ luster
  - ☐ weave ☐ nap
- ☒ Fabric is reversible
- ☒ It looks the same on both sides
- ☐ It stretches easily
- ☐ It will not stretch
- ☒ Fabric tears easily
- ☐ It is difficult to tear
- ☐ Fabric will not tear
- ☐ Pins and needles leave holes, marks
- ☐ It is difficult to ease sleeves and curves
- ☐ It tends to pucker
- ☐ It tends to unravel
- ☐ Inner construction shows from outside
- ☐ Machine eats fabric
- ☐ Skipped stitches
- ☐ Layers feed unevenly
- ☐ Multiple layers are difficult to cut, sew
- ☒ It creases easily
- ☐ Won't hold a crease

Cost per yard
- ☒ Less than $5
- ☒ $5 to $10
- ☐ $10 to $15
- ☐ $15 to $20
- ☐ $20 to $25
- ☐ More than $25

Piece-dyed fabrics
Piece dyeing is the most common way to add color to cotton. A continuous length of fabric is passed through a hot dyebath, then squeezed between padded rollers to even the color and remove any excess liquid. Piece-dyed fabrics usually are solid in color. A pattern can be produced by resist printing, which prevents the dye from penetrating parts of the fabric, or by discharge printing, which removes color from parts of the fabric after it is dyed. Stripes, woven colored patterns and other special effects can be produced by combining fibers that react differently to the same dye.

PIECE-DYED SOLID
YARN-DYED CHECKS

Wearability
- ☒ Durable ☐ Fragile
- ☐ Strong ☐ Weak
- ☒ It is long-wearing
- ☒ It wears evenly
- ☐ It wears out along seams and folds
- ☐ Seams don't hold up under stress
- ☐ Finish wears off
- ☐ Subject to abrasion
- ☒ It resists abrasion
- ☐ Subject to snags
- ☒ It resists snags
- ☐ Subject to runs
- ☐ It tends to pill
- ☐ It tends to shed
- ☐ It produces lint
- ☐ It attracts lint
- ☐ It attract static
- ☐ It tends to cling
- ☒ It holds its shape
- ☐ It loses its shape
- ☐ It stretches out of shape easily
- ☐ It droops, bags
- ☒ It tends to wrinkle
- ☐ It resists wrinkles
- ☐ It crushes easily
- ☐ Water drops leave spots or marks

Suggested care
- ☐ Dry clean only
- ☐ Do not dry clean
- ☐ Dry clean or wash
- ☐ Gently handwash in lukewarm water
- ☐ Roll in a towel to remove moisture
- ☐ Drip dry
- ☐ Lay flat to dry
- ☒ Machine wash
  - ☐ gentle/delicate
  - ☒ regular/normal
- ☒ Machine dry
  - ☐ cool ☒ normal
  - ☐ perm. press
- ☐ Press damp fabric
- ☒ Press dry fabric
- ☐ Dry iron ☒ Steam
- ☐ Iron on wrong side
- ☐ Use a press cloth
- ☐ Use a needleboard
- ☐ Needs no ironing
- ☒ Do not iron
- ☒ Fabric may shrink
- ☐ May bleed or fade
- ☒ Finish washes out

Where to find
- ☒ Any fabric store
- ☒ Major chain store
- ☒ Stores that carry high quality fabric
- ☐ Fabric club
- ☐ Mail order
- ☒ Wholesale supplier

Sweater knits

Thick, heavy sweater knits are hard to find because sweaters are not made the same way as garments from woven goods. The best sweaters are made on a machine that produces shaped pieces of the garment, like hand knitting, rather than yardage. There is very little waste. Less expensive sweaters are cut out and stitched together from knit yardage, just like garments cut from woven fabrics. The cut knits ravel easily and garments have bulky seams. Leftover yardage is sometimes sold to fabric stores.

interlock knit

Fine-gauge, closely knit fabric with indistinct lengthwise ribs on both sides. It is usually made of cotton or a cotton/polyester blend. The fabric is smooth and flat, with a firm hand and a controlled stretch only in the crosswise direction. The best grades are mercerized and made with very fine yarns of combed Pima cotton. The weight varies from light to medium, but interlock is usually thicker, heavier and of better quality than its cousin, jersey. Unlike jersey, it lies flat and does not curl at the edges. It may be plain, printed or striped. Some versions have a touch of spandex for added stretch and control.

How to use

Cotton interlock has a graceful drape that falls into soft flares. It may be gathered into a moderately limp fullness or stretched and molded to the body. Use to make underwear, T-shirts and casual clothing. Interlock is subject to snags, but it wears evenly and is easy to launder. Machine wash and tumble dry, but beware of relaxation shrinkage, especially in the lengthwise direction.

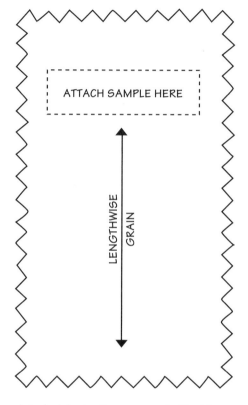

Interlock looks the same on both sides.

Balbriggan
A plain-stitch, tubular knit fabric made of cotton or synthetic fibers or blends, used to make underwear, pajamas and other items. It usually has a napped back and is dyed a deep cream or light tan color. The fabric was first made in Balbriggan, Ireland, and used to make hosiery.

leotard fabric
A general term for cotton/spandex fabrics, including interlock and jersey, that are used to make leotards and exercise wear.

mesh knit
General term for a variety of knit fabrics that have open spaces between the yarns.

stretch interlock
An interlock fabric made with a touch of Lycra® or another spandex fiber, which adds stretch and stretch control. The amount of spandex varies from 5 to 15 percent or more. Cotton/spandex versions are more expensive than all-cotton fabrics. Used for exercise tights and leotards.

Sewing rating
- [ ] Easy to sew
- [ ] Moderately easy
- [x] Average
- [ ] Moderately difficult
- [ ] Extremely difficult

Suggested fit
- [x] Stretch to fit
- [x] Close-fitting
- [ ] Fitted
- [ ] Semi-fitted
- [x] Loose-fitting
- [ ] Very loose-fitting

Suggested styles
- [ ] Pleats [ ] Tucks
  - [ ] pressed
  - [ ] unpressed
- [x] Gathers
  - [ ] limp [x] soft
  - [ ] full [ ] lofty
  - [ ] bouffant
- [x] Elasticized shirring
- [ ] Smocked
- [ ] Tailored
- [ ] Shaped with seams to eliminate bulk
- [ ] Lined
- [ ] Unlined
- [ ] Puffed or bouffant
- [x] Loose and full
- [x] Soft and flowing
- [x] Draped
- [ ] Cut on bias
- [x] Stretch styling

What to expect
- [x] Difficult to cut out
- [ ] Fabric has one-way
  - [ ] design [ ] luster
  - [ ] weave [ ] nap
- [x] Fabric is reversible
- [x] It looks the same on both sides
- [x] It stretches easily
- [ ] It will not stretch
- [ ] Fabric tears easily
- [ ] It is difficult to tear
- [x] Fabric will not tear
- [ ] Pins and needles leave holes, marks
- [ ] It is difficult to ease sleeves and curves
- [ ] It tends to pucker
- [ ] It tends to unravel
- [ ] Inner construction shows from outside
- [ ] Machine eats fabric
- [ ] Skipped stitches
- [x] Layers feed unevenly
- [ ] Multiple layers are difficult to cut, sew
- [ ] It creases easily
- [x] Won't hold a crease

Cost per yard
- [ ] Less than $5
- [x] $5 to $10
- [x] $10 to $15
- [ ] $15 to $20
- [ ] $20 to $25
- [ ] More than $25

Cotton/polyester blends
Cotton is often combined with polyester or other synthetic fibers. Such fabrics are commonly called blends, but they are not all made the same way or for the same reason. A true blend is a mixture of fibers that differ in composition, length, diameter and/or color, spun into the same yarn. A combination yarn has strands of different fibers twisted into a ply yarn. A union fabric has yarns of different fibers in each direction. Blends are used to give the fabric desirable characteristics of each fiber, to reduce costs, to improve spinning and weaving processes and to improve cotton's ability to take finishes, especially durable-press finishes.

Wearability
- [x] Durable [ ] Fragile
- [ ] Strong [ ] Weak
- [x] It is long-wearing
- [x] It wears evenly
- [ ] It wears out along seams and folds
- [ ] Seams don't hold up under stress
- [ ] Finish wears off
- [ ] Subject to abrasion
- [ ] It resists abrasion
- [x] Subject to snags
- [ ] It resists snags
- [x] Subject to runs
- [ ] It tends to pill
- [ ] It tends to shed
- [ ] It produces lint
- [x] It attracts lint
- [ ] It attract static
- [ ] It tends to cling
- [x] It holds its shape
- [ ] It loses its shape
- [ ] It stretches out of shape easily
- [ ] It droops, bags
- [ ] It tends to wrinkle
- [x] It resists wrinkles
- [ ] It crushes easily
- [ ] Water drops leave spots or marks

Suggested care
- [ ] Dry clean only
- [ ] Do not dry clean
- [ ] Dry clean or wash
- [ ] Gently handwash in lukewarm water
- [ ] Roll in a towel to remove moisture
- [ ] Drip dry
- [ ] Lay flat to dry
- [x] Machine wash
  - [x] gentle/delicate
  - [ ] regular/normal
- [x] Machine dry
  - [x] cool [ ] normal
  - [ ] perm. press
- [ ] Press damp fabric
- [ ] Press dry fabric
- [ ] Dry iron [ ] Steam
- [ ] Iron on wrong side
- [ ] Use a press cloth
- [ ] Use a needleboard
- [x] Needs no ironing
- [ ] Do not iron
- [x] Fabric may shrink
- [ ] May bleed or fade
- [ ] Finish washes out

Where to find
- [ ] Any fabric store
- [x] Major chain store
- [x] Stores that carry high quality fabric
- [x] Fabric club
- [x] Mail order
- [x] Wholesale supplier

jersey

A lightweight, smooth, single-knit fabric with flat indistinct ribs on the front and a plain back. The best qualities are made from fine, combed cotton yarns, but the fabric may also be a cotton/polyester blend. Jersey has an easy crosswise stretch, but only a slight lengthwise stretch. Sometimes, the selvages are stiffened with sizing to control jersey's tendency to curl at the edges. The fabric may be plain, printed, napped or knitted with colored yarns to form crosswise stripes. It originally was made of wool on the Jersey and Guernsey Islands, off the coast of England, and used to make heavy fishermen's sweaters.

How to use
Cotton jersey has a limp drape that falls into soft flares. It may be gathered or shirred into a soft fullness or shaped by stretching, especially when the fabric contains a touch of spandex. Use to make loose-fitting or close-fitting T-shirts, exercise wear, underwear, casual dresses and children's clothing. Machine wash and dry. Jersey usualy shrinks the first time it is washed.

ATTACH SAMPLE HERE

LENGTHWISE GRAIN

Jersey has ribs on the face and a plain back.

Spandex
The stretch in stretch jersey is from a rubber-like synthetic fiber called spandex, composed of segmented polyurethane. The segments imitate rubber's natural ability to stretch and recoil, but the synthetic fiber is better: It is stronger and more flexible, it resists abrasion and it is not affected by body acids. Rubber loses its stretch and becomes brittle with age. The best-known spandex fibers are Lycra® by DuPont, Blue C® by Monsanto and Glospan® by Globe Manufacturing.

---

matte jersey
A slinky single-knit fabric with a dull luster, usually made of rayon.

silk jersey
A luxurious, very expensive silk knit used to make long underwear and other clothing. It is sometimes called milanese knit.

stretch jersey
A cotton or synthetic fabric mixed with a small amount of spandex, usually Lycra®, to add stretch and stretch control. It is used to make stretch-to-fit exercise wear and other clothing. The amount of spandex varies from about 5 to 15 percent. Stretch jersey is more expensive than plain jersey.

T-shirt knit
Jersey fabric made in a tube on a circular knitting machine. Used to make T-shirts and other clothing with no side seams.

wool jersey
A warm, comfortable fabric used to make dresses. It is usually heavier than cotton.

Sewing rating
- [ ] Easy to sew
- [ ] Moderately easy
- [x] Average
- [ ] Moderately difficult
- [ ] Extremely difficult

Suggested fit
- [x] Stretch to fit
- [ ] Close-fitting
- [ ] Fitted
- [ ] Semi-fitted
- [x] Loose-fitting
- [x] Very loose-fitting

Suggested styles
- [ ] Pleats [ ] Tucks
  - [ ] pressed
  - [ ] unpressed
- [x] Gathers
  - [x] limp [x] soft
  - [ ] full [ ] lofty
  - [ ] bouffant
- [x] Elasticized shirring
- [ ] Smocked
- [ ] Tailored
- [ ] Shaped with seams to eliminate bulk
- [ ] Lined
- [ ] Unlined
- [ ] Puffed or bouffant
- [x] Loose and full
- [x] Soft and flowing
- [x] Draped
- [ ] Cut on bias
- [x] Stretch styling

What to expect
- [ ] Difficult to cut out
- [ ] Fabric has one-way
  - [ ] design [ ] luster
  - [ ] weave [ ] nap
- [ ] Fabric is reversible
- [ ] It looks the same on both sides
- [x] It stretches easily
- [ ] It will not stretch
- [ ] Fabric tears easily
- [ ] It is difficult to tear
- [x] Fabric will not tear
- [ ] Pins and needles leave holes, marks
- [ ] It is difficult to ease sleeves and curves
- [ ] It tends to pucker
- [ ] It tends to unravel
- [ ] Inner construction shows from outside
- [ ] Machine eats fabric
- [x] Skipped stitches
- [x] Layers feed unevenly
- [ ] Multiple layers are difficult to cut, sew
- [ ] It creases easily
- [x] Won't hold a crease

Cost per yard
- [ ] Less than $5
- [x] $5 to $10
- [x] $10 to $15
- [ ] $15 to $20
- [ ] $20 to $25
- [ ] More than $25

Tube knits
Cotton jerseys and ribbings are often made on a circular knitting machine, which knits a continuous tube of fabric with no selvages. These knits are used to make inexpensive T-shirts and other garments without any side seams, although the fabric may also be cut and stitched in a traditional manner. Tube knits are often made with high-speed machines, which stretch and distort the fabric, causing the finished goods to twist and shrink badly. Some knit fabrics are laced together at the selvages with a loose whipstitch, forming a tube, but this is to minimize distortion and is not a real tube knit.

Wearability
- [x] Durable [ ] Fragile
- [ ] Strong [ ] Weak
- [x] It is long-wearing
- [x] It wears evenly
- [ ] It wears out along seams and folds
- [ ] Seams don't hold up under stress
- [ ] Finish wears off
- [ ] Subject to abrasion
- [ ] It resists abrasion
- [x] Subject to snags
- [ ] It resists snags
- [x] Subject to runs
- [ ] It tends to pill
- [ ] It tends to shed
- [ ] It produces lint
- [ ] It attracts lint
- [ ] It attract static
- [ ] It tends to cling
- [ ] It holds its shape
- [x] It loses its shape
- [x] It stretches out of shape easily
- [ ] It droops, bags
- [ ] It tends to wrinkle
- [x] It resists wrinkles
- [ ] It crushes easily
- [ ] Water drops leave spots or marks

Suggested care
- [ ] Dry clean only
- [ ] Do not dry clean
- [ ] Dry clean or wash
- [x] Gently handwash in lukewarm water
- [x] Roll in a towel to remove moisture
- [ ] Drip dry
- [x] Lay flat to dry
- [x] Machine wash
  - [x] gentle/delicate
  - [ ] regular/normal
- [x] Machine dry
  - [x] cool [ ] normal
  - [ ] perm. press
- [ ] Press damp fabric
- [ ] Press dry fabric
- [ ] Dry iron [ ] Steam
- [ ] Iron on wrong side
- [ ] Use a press cloth
- [ ] Use a needleboard
- [x] Needs no ironing
- [ ] Do not iron
- [x] Fabric may shrink
- [ ] May bleed or fade
- [x] Finish washes out

Where to find
- [x] Any fabric store
- [x] Major chain store
- [x] Stores that carry high quality fabric
- [x] Fabric club
- [x] Mail order
- [x] Wholesale supplier

lawn

A fine, relatively sheer cotton fabric, made with a tight plain weave and a high thread count. Lawn is usually made of fine, combed single yarns, although some less expensive versions are made of carded yarns. The flat smooth fabric may be soft or crisp, depending on the finish, but it is not as soft as batiste or voile, nor as crisp as organdy. Lawn usually is printed with delicate floral patterns, but the fabric may also be bleached or dyed in pastel shades. It may be mercerized to add strength and luster or lightly sized. Lawn gets its name from Laon, France, where it was first made of linen in an open weave.

Liberty prints

Liberty of London is well-known for its beautiful cotton prints. Liberty uses copper rollers to print exquisite patterns with very fine lines, rather than the more typical rubber or wooden rollers. Arthur Liberty opened his first store, called East India House, in 1875 and sold fabrics and goods imported from India, China, Japan and Persia. Demand soon exceeded supply, so Liberty imported plain woven goods and began to print his own. The first prints were exact copies of old Indian prints.

How to use

Cotton lawn has a graceful drape that falls into delicate flares. It may be gathered, smocked or shirred into a soft, limp fullness. It is moderately easy to cut and sew. Use to make semi-fitted, loose or very loose-fitting blouses, dresses, lingerie and baby clothes. It is also used to make collars, cuffs and handkerchiefs. Machine wash and dry on delicate cycles. Iron with light steam.

ATTACH SAMPLE HERE

LENGTHWISE GRAIN

Lawn printed with a delicate floral design.

bishop's lawn
A light, fine cotton lawn that is usually white or printed and given a bluish starch finish. Originally made in England and once in great demand. Also called Victoria lawn.

linen lawn
Fine linen fabric made with an open weave.

longcloth
A fine, soft, bleached cotton fabric made with a tight plain weave and loosely twisted combed or carded yarns. It is finished with little or no sizing. Longcloth is heavier than lawn and more closely woven. It was one of the first fabrics woven in long pieces, hence its name.

Tana Lawn®
Registered trademark for lawn prints by Liberty of London, named for Lake Tana in the Sudan where the raw cotton is grown.

Victoria lawn
A closely woven, plain cotton lawn, usually about 38 inches wide and made in England.

Sewing rating
- [ ] Easy to sew
- [x] Moderately easy
- [ ] Average
- [ ] Moderately difficult
- [ ] Extremely difficult

Suggested fit
- [ ] Stretch to fit
- [ ] Close-fitting
- [x] Fitted
- [x] Semi-fitted
- [x] Loose-fitting
- [ ] Very loose-fitting

Suggested styles
- [x] Pleats [x] Tucks
  - [x] pressed
  - [ ] unpressed
- [x] Gathers
  - [x] limp [x] soft
  - [ ] full [ ] lofty
  - [ ] bouffant
- [x] Elasticized shirring
- [x] Smocked
- [ ] Tailored
- [ ] Shaped with seams to eliminate bulk
- [ ] Lined
- [ ] Unlined
- [ ] Puffed or bouffant
- [x] Loose and full
- [x] Soft and flowing
- [x] Draped
- [ ] Cut on bias
- [ ] Stretch styling

What to expect
- [ ] Difficult to cut out
- [ ] Fabric has one-way
  - [ ] design [ ] luster
  - [ ] weave [ ] nap
- [ ] Fabric is reversible
- [ ] It looks the same on both sides
- [ ] It stretches easily
- [ ] It will not stretch
- [x] Fabric tears easily
- [ ] It is difficult to tear
- [ ] Fabric will not tear
- [x] Pins and needles leave holes, marks
- [x] It is difficult to ease sleeves and curves
- [ ] It tends to pucker
- [ ] It tends to unravel
- [x] Inner construction shows from outside
- [x] Machine eats fabric
- [ ] Skipped stitches
- [ ] Layers feed unevenly
- [ ] Multiple layers are difficult to cut, sew
- [x] It creases easily
- [ ] Won't hold a crease

Cost per yard
- [ ] Less than $5
- [x] $5 to $10
- [x] $10 to $15
- [x] $15 to $20
- [x] $20 to $25
- [ ] More than $25

Egyptian cotton
Liberty of London uses cotton from the Sudan to make its fine fabrics, and with good reason: Egypt produces some of the finest cotton in the world. The fertile soil of the Nile Delta and warm climate provide excellent conditions for growing extra-long-staple cotton. Egypt, in fact, is the world's biggest producer of long-staple cotton. It has been growing cotton commercially since about 1820, when a French engineer visiting Cairo noticed a tree cotton with fibers superior to the coarse native cottons. The tree cotton was mixed with Sea Island and South American cottons to produce today's prized hybrid, which is picked by hand to preserve the quality.

Wearability
- [ ] Durable [ ] Fragile
- [ ] Strong [ ] Weak
- [x] It is long-wearing
- [x] It wears evenly
- [ ] It wears out along seams and folds
- [ ] Seams don't hold up under stress
- [ ] Finish wears off
- [ ] Subject to abrasion
- [ ] It resists abrasion
- [ ] Subject to snags
- [x] It resists snags
- [ ] Subject to runs
- [ ] It tends to pill
- [ ] It tends to shed
- [ ] It produces lint
- [ ] It attracts lint
- [ ] It attract static
- [x] It tends to cling
- [x] It holds its shape
- [ ] It loses its shape
- [ ] It stretches out of shape easily
- [ ] It droops, bags
- [x] It tends to wrinkle
- [ ] It resists wrinkles
- [ ] It crushes easily
- [x] Water drops leave spots or marks

Suggested care
- [ ] Dry clean only
- [ ] Do not dry clean
- [ ] Dry clean or wash
- [x] Gently handwash in lukewarm water
- [x] Roll in a towel to remove moisture
- [x] Drip dry
- [ ] Lay flat to dry
- [x] Machine wash
  - [x] gentle/delicate
  - [ ] regular/normal
- [x] Machine dry
  - [x] cool [ ] normal
  - [ ] perm. press
- [ ] Press damp fabric
- [x] Press dry fabric
- [ ] Dry iron [x] Steam
- [ ] Iron on wrong side
- [ ] Use a press cloth
- [ ] Use a needleboard
- [ ] Needs no ironing
- [ ] Do not iron
- [x] Fabric may shrink
- [ ] May bleed or fade
- [x] Finish washes out

Where to find
- [ ] Any fabric store
- [x] Major chain store
- [x] Stores that carry high quality fabric
- [x] Fabric club
- [x] Mail order
- [x] Wholesale supplier

Madras

A soft cotton fabric with a plaid, checked or striped pattern, handwoven in the village cottages of Madras, India. True Madras is woven with colored yarns that have been dyed with vegetable dyes. When the fabric is washed, the dyes bleed together into lovely, muted colors. Weavers keep their spools and spindles wet to encourage bleeding while the fabric is still on the loom. Slubbed yarns and imperfections are normal and the fabric sometimes has an odd, swampy smell from being washed many times to promote bleeding. Madras is widely imitated — the imitations are supposed to be clearly marked.

How to use
Madras has a soft drape that falls into moderately soft flares. It may be pleated, gathered or shirred into a soft fullness. Yardage requirements may need adjusting to accommodate its narrower width. Use to make semi-fitted or loose-fitting blouses, shirts and dresses. Madras shrinks badly the first time it is washed and continues to bleed and fade each time it is washed.

Authentic, handwoven Indian Madras.

Vegetable dyes
Roots, bark, berries and other plant matter are the sources of some of the oldest known dyes, called vegetable dyes. Madder (red) and indigo (blue) are the best known. Other colors come from oak bark (brown), myrtle bushes (gray), cocklebur leaves (yellow), sumac berries (black), sassafrass (orange), walnut hulls (brown), hickory (green) and poke berries (red). Today, most vegetable dyes have been replaced by synthetic dyes, which are easier to control and less likely to bleed and fade.

imitation madras
A muted plaid fabric woven by machine with colored yarns, in imitation of true Madras. Some versions are printed.

Madras cotton
A type of cotton grown in Madras, India. The fiber contains a large quantity of dirt.

Madras gauze
A very lightweight gauze decorated with patterns formed by coarse, extra filling yarns. Floats on the back are cut away.

Madras gingham
Fine cotton fabric that is lighter in weight and more colorful than regular Madras.

plaid
A term used to describe any fabric with a pattern of multi-colored stripes or bars that cross each other at right angles to form squares or rectangles. The pattern is usually woven with colored yarns, but some versions are printed. When only two colors are used, it is called check or gingham.

Sewing rating
- ☐ Easy to sew
- ☐ Moderately easy
- ☒ Average
- ☐ Moderately difficult
- ☐ Extremely difficult

Suggested fit
- ☐ Stretch to fit
- ☐ Close-fitting
- ☐ Fitted
- ☒ Semi-fitted
- ☒ Loose-fitting
- ☒ Very loose-fitting

Suggested styles
- ☒ Pleats ☒ Tucks
  - ☒ pressed
  - ☐ unpressed
- ☒ Gathers
  - ☐ limp ☒ soft
  - ☐ full ☐ lofty
  - ☐ bouffant
- ☒ Elasticized shirring
- ☐ Smocked
- ☐ Tailored
- ☐ Shaped with seams to eliminate bulk
- ☐ Lined
- ☐ Unlined
- ☐ Puffed or bouffant
- ☒ Loose and full
- ☒ Soft and flowing
- ☐ Draped
- ☐ Cut on bias
- ☐ Stretch styling

What to expect
- ☐ Difficult to cut out
- ☒ Fabric has one-way
  - ☒ design ☐ luster
  - ☐ weave ☐ nap
- ☒ Fabric is reversible
- ☒ It looks the same on both sides
- ☐ It stretches easily
- ☐ It will not stretch
- ☒ Fabric tears easily
- ☐ It is difficult to tear
- ☐ Fabric will not tear
- ☐ Pins and needles leave holes, marks
- ☐ It is difficult to ease sleeves and curves
- ☐ It tends to pucker
- ☐ It tends to unravel
- ☐ Inner construction shows from outside
- ☐ Machine eats fabric
- ☐ Skipped stitches
- ☐ Layers feed unevenly
- ☐ Multiple layers are difficult to cut, sew
- ☒ It creases easily
- ☐ Won't hold a crease

Cost per yard
- ☒ Less than $5
- ☒ $5 to $10
- ☐ $10 to $15
- ☐ $15 to $20
- ☐ $20 to $25
- ☐ More than $25

Cotton from India
The Madras area in southeastern India has a highly developed cotton industry and is the source of many hand-loomed fabrics. The industry flourishes all over India because the cotton plant grows easily in the warm, dry climate. Indian cotton is usually named for the region where it is grown. The better varieties include Bengal, Oomra, Punjab, Cambodia and Broach. The finest Indian cotton once came from the Dacca area (the region is now part of Bangladesh). For centuries, India ranked first in the production of cotton, but it now falls behind China and the United States. Most Indian cotton has shorter, coarser fibers than U.S. cotton.

Wearability
- ☐ Durable ☐ Fragile
- ☐ Strong ☒ Weak
- ☐ It is long-wearing
- ☒ It wears evenly
- ☐ It wears out along seams and folds
- ☒ Seams don't hold up under stress
- ☐ Finish wears off
- ☐ Subject to abrasion
- ☐ It resists abrasion
- ☐ Subject to snags
- ☐ It resists snags
- ☐ Subject to runs
- ☐ It tends to pill
- ☐ It tends to shed
- ☐ It produces lint
- ☐ It attracts lint
- ☐ It attract static
- ☐ It tends to cling
- ☒ It holds its shape
- ☐ It loses its shape
- ☐ It stretches out of shape easily
- ☐ It droops, bags
- ☒ It tends to wrinkle
- ☐ It resists wrinkles
- ☐ It crushes easily
- ☐ Water drops leave spots or marks

Suggested care
- ☐ Dry clean only
- ☐ Do not dry clean
- ☐ Dry clean or wash
- ☐ Gently handwash in lukewarm water
- ☐ Roll in a towel to remove moisture
- ☐ Drip dry
- ☐ Lay flat to dry
- ☒ Machine wash
  - ☒ gentle/delicate
  - ☐ regular/normal
- ☒ Machine dry
  - ☒ cool ☐ normal
  - ☐ perm. press
- ☒ Press damp fabric
- ☐ Press dry fabric
- ☒ Dry iron ☐ Steam
- ☐ Iron on wrong side
- ☐ Use a press cloth
- ☐ Use a needleboard
- ☐ Needs no ironing
- ☐ Do not iron
- ☒ Fabric may shrink
- ☒ May bleed or fade
- ☐ Finish washes out

Where to find
- ☐ Any fabric store
- ☒ Major chain store
- ☒ Stores that carry high quality fabric
- ☐ Fabric club
- ☒ Mail order
- ☒ Wholesale supplier

Unstable weaves

Like most loosely woven fabrics, monk's cloth is not very stable. The fabric is highly susceptible to yarn slippage, a term used to describe yarns that slip out of place, leaving unsightly open spaces in the fabric. Unstable weaves also tend to stretch, droop, bag or sag. The problem is exaggerated when the fabric must support a great deal of its own weight, as is the case with draperies. A tightly woven, stable underlining may be used to help the fabric maintain its shape and to control slippage.

monk's cloth

A thick, soft, coarse cotton cloth made with a loose basket weave and softly spun, carded ply yarns, sometimes mixed with a small amount of flax or jute. The cloth is usually made with a 2x2 or 4x4 weave, but 6x6 and 8x8 also are used. The medium to heavy fabric has the same number of yarns per inch in each direction. Monk's cloth is usually a characteristic natural oatmeal color, but it may also be dyed, bleached white or woven in solid colors, stripes or plaids. It is also called friar's cloth, abbot's cloth, druid's cloth, bishop's cloth, belfry cloth and mission cloth, some of which are registered trade names.

How to use

Monk's cloth has a soft, full drape that falls into moderately soft flares. It may be gathered into a soft fullness. The loose weave snags easily, ravels badly and is prone to bagging and sagging. The fabric also shrinks badly and is subject to yarn and seam slippage. Use to make draperies, slip covers and upholstery. Dry clean or launder gently at home, but preshrink before using.

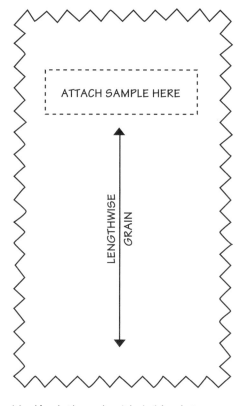

Monk's cloth made with 4x4 basket weave.

---

cloister cloth
A rough cotton drapery fabric made with a basket weave. It is not as heavy as monk's cloth and less bulky.

fancy monk's cloth
A heavy, absorbent cotton toweling made with ply yarns and a honeycomb weave.

hopsacking
A soft, loosely woven cotton fabric made with a plain or basket weave and thick, coarse ply yarns of low-grade cotton. It is used to make dresses, coats and when printed, decorative fabrics. The fabric gets its name from the burlap sacks used to carry hops from the fields to the brewery.

matt weave
Another name for the basket weave.

sack cloth
A term used to describe a coarse fabric of cotton or linen, usually worn as a sign of mourning or penance. There are a number of references to sack cloth in the Bible.

Sewing rating
- [ ] Easy to sew
- [ ] Moderately easy
- [x] Average
- [ ] Moderately difficult
- [ ] Extremely difficult

Suggested fit
- [ ] Stretch to fit
- [ ] Close-fitting
- [x] Fitted
- [x] Semi-fitted
- [x] Loose-fitting
- [ ] Very loose-fitting

Suggested styles
- [x] Pleats [x] Tucks
    - [ ] pressed
    - [x] unpressed
- [x] Gathers
    - [ ] limp [ ] soft
    - [x] full [ ] lofty
    - [ ] bouffant
- [ ] Elasticized shirring
- [ ] Smocked
- [ ] Tailored
- [x] Shaped with seams to eliminate bulk
- [x] Lined
- [ ] Unlined
- [ ] Puffed or bouffant
- [x] Loose and full
- [ ] Soft and flowing
- [x] Draped
- [ ] Cut on bias
- [ ] Stretch styling

What to expect
- [ ] Difficult to cut out
- [ ] Fabric has one-way
    - [ ] design [ ] luster
    - [ ] weave [ ] nap
- [x] Fabric is reversible
- [x] It looks the same on both sides
- [x] It stretches easily
- [ ] It will not stretch
- [ ] Fabric tears easily
- [ ] It is difficult to tear
- [x] Fabric will not tear
- [ ] Pins and needles leave holes, marks
- [ ] It is difficult to ease sleeves and curves
- [ ] It tends to pucker
- [x] It tends to unravel
- [ ] Inner construction shows from outside
- [ ] Machine eats fabric
- [ ] Skipped stitches
- [ ] Layers feed unevenly
- [x] Multiple layers are difficult to cut, sew
- [ ] It creases easily
- [x] Won't hold a crease

Cost per yard
- [ ] Less than $5
- [x] $5 to $10
- [ ] $10 to $15
- [ ] $15 to $20
- [ ] $20 to $25
- [ ] More than $25

The basket weave
The basket weave is a variation of the plain weave, made by grouping yarns and weaving them as one to form a checkered pattern. The simplest version is called a 2x2 basket weave, made by weaving pairs of yarns as single yarns in both directions. A semi-basket weave has yarns that are grouped in only one direction (2x1). A plain basket weave has square blocks; a fancy basket weave has oblong blocks. Fabrics often are loosely woven and tend to resist wrinkles. Monk's cloth, homespun, hopsacking and oxford cloth are examples of the basket weave.

Wearability
- [ ] Durable [ ] Fragile
- [x] Strong [ ] Weak
- [ ] It is long-wearing
- [ ] It wears evenly
- [ ] It wears out along seams and folds
- [x] Seams don't hold up under stress
- [ ] Finish wears off
- [x] Subject to abrasion
- [ ] It resists abrasion
- [x] Subject to snags
- [ ] It resists snags
- [ ] Subject to runs
- [ ] It tends to pill
- [ ] It tends to shed
- [x] It produces lint
- [ ] It attracts lint
- [ ] It attract static
- [ ] It tends to cling
- [ ] It holds its shape
- [x] It loses its shape
- [x] It stretches out of shape easily
- [x] It droops, bags
- [ ] It tends to wrinkle
- [x] It resists wrinkles
- [ ] It crushes easily
- [ ] Water drops leave spots or marks

Suggested care
- [ ] Dry clean only
- [ ] Do not dry clean
- [x] Dry clean or wash
- [ ] Gently handwash in lukewarm water
- [ ] Roll in a towel to remove moisture
- [ ] Drip dry
- [ ] Lay flat to dry
- [x] Machine wash
    - [x] gentle/delicate
    - [ ] regular/normal
- [x] Machine dry
    - [ ] cool [x] normal
    - [ ] perm. press
- [ ] Press damp fabric
- [ ] Press dry fabric
- [ ] Dry iron [ ] Steam
- [ ] Iron on wrong side
- [ ] Use a press cloth
- [ ] Use a needleboard
- [x] Needs no ironing
- [ ] Do not iron
- [x] Fabric may shrink
- [ ] May bleed or fade
- [ ] Finish washes out

Where to find
- [ ] Any fabric store
- [ ] Major chain store
- [x] Stores that carry high quality fabric
- [ ] Fabric club
- [ ] Mail order
- [x] Wholesale supplier

muslin

General term for large group of inexpensive, plain-weave fabrics made of cotton or a cotton/polyester blend and a thread count of up to about 160 yarns per square inch. The weight and quality vary considerably, from light sheers to heavy sheeting, but the typical muslin is of average weight and quality with a natural, unbleached color and finished with light or heavy sizing. Sometimes it is bleached white, but when printed or dyed, it is called by other names. It is named for Mosul, Mesopotamia (now Iraq), where it was first made. During the Middle Ages, the term referred to heavy, coarse fabrics. Muslin as we know it today was first made in Europe in about 1700.

How to use
Muslin has a moderately limp to moderately stiff drape that can be pleated, gathered, smocked or shirred. It is used most often to make press cloths and sample garments, but it may also be used for underwear, aprons, linings, shirts, dresses, sheets, pillowcases and furniture coverings. Machine wash.

Unbleached muslin is natural in color.

Sizing
Sizing is the general term for a starch, wax, gelatin, oil or other stiffener that may be applied to the yarn or fabric. It is used to increase smoothness, weight, strength, abrasion resistance, luster or stiffness. In general, good quality fabrics have little or no sizing, while inexpensive cottons often are heavily sized, making them appear firmer and sturdier. Heavily sized fabrics may appear to be coated with a filmy or powdery substance. Sizing usually washes out and the fabric becomes softer.

---

double-wide cloth
Label given to muslin, percale and sheeting fabrics that are 90 inches wide instead of the usual 45 inches. The term usually is shortened to double cloth, but it shouldn't be confused with true double cloth, a fabric with two distinct layers that are woven together as one cloth.

muslin sheeting
Firmly woven fabric with strong taped selvages, made of carded cotton in several versions of the plain weave, the most common of which are 56x56, 64x64 and 72x68. The fabric usually is bleached and finished with a small amount of sizing. It is used to make sheets and pillowcases.

percale
Smooth sheeting fabric that is finer and more tightly woven than muslin, usually made with combed yarns and a balanced plain weave. It has a thread count of 160 or more per square inch. Percale usually is mercerized to add strength and may be dyed, bleached white or printed.

Sewing rating
- ☒ Easy to sew
- ☐ Moderately easy
- ☐ Average
- ☐ Moderately difficult
- ☐ Extremely difficult

Suggested fit
- ☐ Stretch to fit
- ☒ Close-fitting
- ☒ Fitted
- ☒ Semi-fitted
- ☐ Loose-fitting
- ☐ Very loose-fitting

Suggested styles
- ☒ Pleats ☒ Tucks
  - ☒ pressed
  - ☐ unpressed
- ☒ Gathers
  - ☐ limp ☒ soft
  - ☒ full ☐ lofty
  - ☐ bouffant
- ☒ Elasticized shirring
- ☒ Smocked
- ☐ Tailored
- ☐ Shaped with seams to eliminate bulk
- ☐ Lined
- ☐ Unlined
- ☐ Puffed or bouffant
- ☒ Loose and full
- ☒ Soft and flowing
- ☒ Draped
- ☐ Cut on bias
- ☐ Stretch styling

What to expect
- ☐ Difficult to cut out
- ☐ Fabric has one-way
  - ☐ design ☐ luster
  - ☐ weave ☐ nap
- ☒ Fabric is reversible
- ☒ It looks the same on both sides
- ☐ It stretches easily
- ☐ It will not stretch
- ☒ Fabric tears easily
- ☐ It is difficult to tear
- ☐ Fabric will not tear
- ☐ Pins and needles leave holes, marks
- ☐ It is difficult to ease sleeves and curves
- ☐ It tends to pucker
- ☐ It tends to unravel
- ☐ Inner construction shows from outside
- ☐ Machine eats fabric
- ☐ Skipped stitches
- ☐ Layers feed unevenly
- ☐ Multiple layers are difficult to cut, sew
- ☒ It creases easily
- ☐ Won't hold a crease

Cost per yard
- ☒ Less than $5
- ☒ $5 to $10
- ☐ $10 to $15
- ☐ $15 to $20
- ☐ $20 to $25
- ☐ More than $25

Thread count
Thread count is used to define the quality of a fabric in terms of threads, or yarns, per square inch. The higher the number, the better the quality. Thread count can be recorded several ways. For example, "96 x 78" means the fabric has 96 warp yarns and 78 filling yarns per inch, while "72 square" means the fabric has 72 yarns per inch in each direction. If a single number is given, such as "200 count," it means the fabric has 200 total yarns per inch, adding the yarns in both directions. This method is used to describe sheets. A thread count of 200 or more is a sign of fine yarns – usually of combed cotton – and high quality.

ONE SQUARE INCH

Wearability
- ☒ Durable ☐ Fragile
- ☐ Strong ☐ Weak
- ☒ It is long-wearing
- ☒ It wears evenly
- ☐ It wears out along seams and folds
- ☐ Seams don't hold up under stress
- ☐ Finish wears off
- ☐ Subject to abrasion
- ☒ It resists abrasion
- ☐ Subject to snags
- ☒ It resists snags
- ☐ Subject to runs
- ☐ It tends to pill
- ☐ It tends to shed
- ☐ It produces lint
- ☐ It attracts lint
- ☐ It attract static
- ☐ It tends to cling
- ☒ It holds its shape
- ☐ It loses its shape
- ☐ It stretches out of shape easily
- ☐ It droops, bags
- ☒ It tends to wrinkle
- ☐ It resists wrinkles
- ☐ It crushes easily
- ☐ Water drops leave spots or marks

Suggested care
- ☐ Dry clean only
- ☐ Do not dry clean
- ☐ Dry clean or wash
- ☐ Gently handwash in lukewarm water
- ☐ Roll in a towel to remove moisture
- ☐ Drip dry
- ☐ Lay flat to dry
- ☒ Machine wash
  - ☐ gentle/delicate
  - ☒ regular/normal
- ☒ Machine dry
  - ☐ cool ☒ normal
  - ☐ perm. press
- ☐ Press damp fabric
- ☒ Press dry fabric
- ☐ Dry iron ☒ Steam
- ☐ Iron on wrong side
- ☐ Use a press cloth
- ☐ Use a needleboard
- ☐ Needs no ironing
- ☒ Do not iron
- ☒ Fabric may shrink
- ☐ May bleed or fade
- ☒ Finish washes out

Where to find
- ☒ Any fabric store
- ☒ Major chain store
- ☒ Stores that carry high quality fabric
- ☒ Fabric club
- ☒ Mail order
- ☒ Wholesale supplier

Organza

Organdy often is confused with organza, but the terms refer to different fabrics. Organza is the grown-up version of organdy, made of silk with a loose plain weave and tightly twisted yarns. The sheer, crisp fabric is not as wiry as organdy and has a soft, pearl-like luster. It is used to make dramatically bouffant evening wear, especially in the form of see-through puffed sleeves and full skirts on willowy models. It also is used to make invisible facings and lightweight linings in silk garments.

organdy

Sheer, lightweight cotton fabric made with an open plain weave and very fine, tightly twisted single yarns. The flat smooth fabric is either very expensive or very inexpensive. The best organdy is almost transparent and feels more like a fine wire screen than a fabric. It is made with combed cotton yarns and treated with a permanently stiff finish. Inexpensive versions are more opaque, slightly heavier and have a crisp starched finish that eventually washes out. Organdy is usually white, but it may be dyed or printed. Some versions are made with a cotton/polyester blend. Similar fabrics are made of rayon and silk.

How to use

Organdy has a crisp to stiff drape that may be gathered, pleated or shirred into a bouffant fullness. It may be used alone or paired with another fabric to add fullness or loft. Use to make semi-fitted, puffed or bouffant evening wear, dresses, blouses, children's clothing, curtains, interfacings and facings. The fabric wrinkles and crushes easily, but it is quickly smoothed with a hot iron.

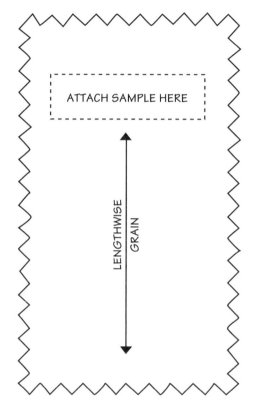

Swiss organdy is stiff and transparent.

organdy finish
A crisp finish applied to fine, sheer fabrics of cotton, rayon, silk and other fibers. It is sometimes permanent and sometimes not.

parchmentized fabric
General term for any type of fabric that has received a Swiss finish.

Swiss finish
Name for a type of parchmentized finish that originated in Switzerland and is now used in other countries as well. The fabric is mercerized before and after the acid treatment to increase its strength and improve the transparent effect. Treated fabrics are costly.

Swiss organdy
A very crisp, transparent organdy made in Switzerland of fine combed-cotton yarns. It feels more like a fine wire screen than a fabric. The permanent finish needs minimal ironing to smooth out wrinkles and restore crispness. The expensive fabric can be found at stores that carry better quality goods.

Sewing rating
- ☐ Easy to sew
- ☐ Moderately easy
- ☒ Average
- ☐ Moderately difficult
- ☐ Extremely difficult

Suggested fit
- ☐ Stretch to fit
- ☒ Close-fitting
- ☒ Fitted
- ☒ Semi-fitted
- ☐ Loose-fitting
- ☐ Very loose-fitting

Suggested styles
- ☒ Pleats ☐ Tucks
  - ☒ pressed
  - ☐ unpressed
- ☒ Gathers
  - ☐ limp ☐ soft
  - ☐ full ☒ lofty
  - ☒ bouffant
- ☒ Elasticized shirring
- ☒ Smocked
- ☐ Tailored
- ☐ Shaped with seams to eliminate bulk
- ☒ Lined
- ☐ Unlined
- ☒ Puffed or bouffant
- ☐ Loose and full
- ☐ Soft and flowing
- ☐ Draped
- ☐ Cut on bias
- ☐ Stretch styling

What to expect
- ☐ Difficult to cut out
- ☐ Fabric has one-way
  - ☐ design ☐ luster
  - ☐ weave ☐ nap
- ☒ Fabric is reversible
- ☒ It looks the same on both sides
- ☐ It stretches easily
- ☐ It will not stretch
- ☒ Fabric tears easily
- ☐ It is difficult to tear
- ☐ Fabric will not tear
- ☒ Pins and needles leave holes, marks
- ☒ It is difficult to ease sleeves and curves
- ☐ It tends to pucker
- ☐ It tends to unravel
- ☒ Inner construction shows from outside
- ☐ Machine eats fabric
- ☐ Skipped stitches
- ☐ Layers feed unevenly
- ☐ Multiple layers are difficult to cut, sew
- ☒ It creases easily
- ☐ Won't hold a crease

Cost per yard
- ☒ Less than $5
- ☒ $5 to $10
- ☐ $10 to $15
- ☐ $15 to $20
- ☒ $20 to $25
- ☒ More than $25

Stiff or crisp finishes
Inexpensive organdy is stiffened with starch or sizing, both of which will come out in the wash. A more durable stiffness is produced by applying a resin finish, but it won't last forever. The finest organdy is made permanently crisp by taking advantage of one of cotton's weaknesses – it dissolves in strong acid. **Parchmentizing** uses sulfuric acid to partially dissolve the fabric, which then is allowed to harden, leaving a thin, wiry fabric that is more transparent and permanently crisp. The process takes great skill, so fabrics are costly. Organdy can be laundered at home and crispness is easily restored with an iron.

Wearability
- ☒ Durable ☐ Fragile
- ☐ Strong ☐ Weak
- ☒ It is long-wearing
- ☒ It wears evenly
- ☐ It wears out along seams and folds
- ☐ Seams don't hold up under stress
- ☐ Finish wears off
- ☐ Subject to abrasion
- ☐ It resists abrasion
- ☐ Subject to snags
- ☒ It resists snags
- ☐ Subject to runs
- ☐ It tends to pill
- ☐ It tends to shed
- ☐ It produces lint
- ☐ It attracts lint
- ☐ It attract static
- ☐ It tends to cling
- ☒ It holds its shape
- ☐ It loses its shape
- ☐ It stretches out of shape easily
- ☐ It droops, bags
- ☒ It tends to wrinkle
- ☐ It resists wrinkles
- ☒ It crushes easily
- ☒ Water drops leave spots or marks

Suggested care
- ☐ Dry clean only
- ☐ Do not dry clean
- ☒ Dry clean or wash
- ☒ Gently handwash in lukewarm water
- ☒ Roll in a towel to remove moisture
- ☒ Drip dry
- ☐ Lay flat to dry
- ☒ Machine wash
  - ☒ gentle/delicate
  - ☐ regular/normal
- ☒ Machine dry
  - ☒ cool ☐ normal
  - ☐ perm. press
- ☒ Press damp fabric
- ☐ Press dry fabric
- ☒ Dry iron ☐ Steam
- ☐ Iron on wrong side
- ☐ Use a press cloth
- ☐ Use a needleboard
- ☐ Needs no ironing
- ☐ Do not iron
- ☐ Fabric may shrink
- ☐ May bleed or fade
- ☒ Finish washes out

Where to find
- ☐ Any fabric store
- ☒ Major chain store
- ☒ Stores that carry high quality fabric
- ☐ Fabric club
- ☒ Mail order
- ☒ Wholesale supplier

osnaburg

A coarse, strong cotton fabric made with a plain weave. The medium to medium-heavy fabric is similar to sheeting, but made with coarser yarns and a looser weave. It is sometimes classified as coarse sheeting. There are two grades of osnaburg: "clean," made of low-grade, all-white cotton, and "part waste," or "P.W.", made with yarns that contain cotton waste. Osnaburg is usually sold as unfinished greige goods, but it may also be printed with stripes, checks or novelty effects and treated to minimize shrinkage. The fabric is named for the city of Osnaburg, Germany, where it was first made of flax and tow, with blue or brown and white yarns that formed stripes, checks or solid colors.

How to use
Osnaburg has a firm drape that falls into wide flares and maintains a lofty fullness. It may be used to make work clothes, box spring covers, mattress ticking, decorative fabrics, slip covers and toweling. The utility fabric is strong and durable, but it shrinks badly the first time it is washed.

Osnaburg is coarse and unfinished.

Greige goods
When fabric comes off the loom, it is in a rough unfinished state, called "in the greige" or "greige goods" by the textile industry. Greige sounds like gray, and it is sometimes spelled that way, but color has nothing to do with it – greige is French for natural. Most greige fabrics are finished by a converter before they are sold to the manufacturer or consumer. One greige fabric can be finished many different ways to produce a variety of fabrics with different characteristics and appearances.

---

converter
Industry term for a person or organization who buys greige goods and arranges to have them finished by bleaching, dyeing, printing, mercerizing and so forth, before selling the goods to apparel manufacturers, wholesalers, retailers and others. The finished cloth is called converted goods.

genetically engineered cotton
Since the early 1990s, scientists have been working on strains of cotton that have been genetically altered to resist disease, insects and weed killers, among other things, in order to decrease the use of fertilizers, pesticides, defoliants and dyes. A naturally colored cotton is in the works.

organic cotton
Cotton grown without chemical fertilizers, pesticides or defoliants.

P.F.P.
Prepared for printing, painting or dyeing. Fabric has either never had a finish applied to it or has had all the finishes washed out.

Sewing rating
- ☒ Easy to sew
- ☐ Moderately easy
- ☐ Average
- ☐ Moderately difficult
- ☐ Extremely difficult

Suggested fit
- ☐ Stretch to fit
- ☐ Close-fitting
- ☒ Fitted
- ☒ Semi-fitted
- ☒ Loose-fitting
- ☐ Very loose-fitting

Suggested styles
- ☒ Pleats ☐ Tucks
  - ☒ pressed
  - ☐ unpressed
- ☒ Gathers
  - ☐ limp ☐ soft
  - ☒ full ☒ lofty
  - ☐ bouffant
- ☐ Elasticized shirring
- ☐ Smocked
- ☐ Tailored
- ☒ Shaped with seams to eliminate bulk
- ☐ Lined
- ☐ Unlined
- ☐ Puffed or bouffant
- ☐ Loose and full
- ☐ Soft and flowing
- ☐ Draped
- ☐ Cut on bias
- ☐ Stretch styling

What to expect
- ☐ Difficult to cut out
- ☐ Fabric has one-way
  - ☐ design ☐ luster
  - ☐ weave ☐ nap
- ☒ Fabric is reversible
- ☒ It looks the same on both sides
- ☐ It stretches easily
- ☐ It will not stretch
- ☐ Fabric tears easily
- ☒ It is difficult to tear
- ☐ Fabric will not tear
- ☐ Pins and needles leave holes, marks
- ☐ It is difficult to ease sleeves and curves
- ☐ It tends to pucker
- ☐ It tends to unravel
- ☐ Inner construction shows from outside
- ☐ Machine eats fabric
- ☐ Skipped stitches
- ☐ Layers feed unevenly
- ☒ Multiple layers are difficult to cut, sew
- ☒ It creases easily
- ☐ Won't hold a crease

Cost per yard
- ☒ Less than $5
- ☐ $5 to $10
- ☐ $10 to $15
- ☐ $15 to $20
- ☐ $20 to $25
- ☐ More than $25

Colored cotton
The quality of a cotton crop is determined in part by its color – and white has always been the most desirable. But today, some growers are developing colored cottons with fibers that are naturally tan, green, brown, yellow, copper, blue and pink. The best known breeder of colored cottons is Arizona's Sally Fox, who sells organically grown green and brown fibers to clothing manufacturers. Fox claims that clothing made from her Fox Fibre® improves with age because the color intensifies in the wash, rather than fading. Another plus: it eliminates the need for dyes, which is better for the environment and more economical to produce.

Wearability
- ☒ Durable ☐ Fragile
- ☒ Strong ☐ Weak
- ☒ It is long-wearing
- ☒ It wears evenly
- ☐ It wears out along seams and folds
- ☐ Seams don't hold up under stress
- ☐ Finish wears off
- ☐ Subject to abrasion
- ☐ It resists abrasion
- ☐ Subject to snags
- ☒ It resists snags
- ☐ Subject to runs
- ☐ It tends to pill
- ☐ It tends to shed
- ☐ It produces lint
- ☐ It attracts lint
- ☐ It attract static
- ☐ It tends to cling
- ☒ It holds its shape
- ☐ It loses its shape
- ☐ It stretches out of shape easily
- ☐ It droops, bags
- ☐ It tends to wrinkle
- ☒ It resists wrinkles
- ☐ It crushes easily
- ☐ Water drops leave spots or marks

Suggested care
- ☐ Dry clean only
- ☐ Do not dry clean
- ☐ Dry clean or wash
- ☐ Gently handwash in lukewarm water
- ☐ Roll in a towel to remove moisture
- ☐ Drip dry
- ☐ Lay flat to dry
- ☒ Machine wash
  - ☐ gentle/delicate
  - ☒ regular/normal
- ☒ Machine dry
  - ☐ cool ☒ normal
  - ☐ perm. press
- ☐ Press damp fabric
- ☐ Press dry fabric
- ☐ Dry iron ☐ Steam
- ☐ Iron on wrong side
- ☐ Use a press cloth
- ☐ Use a needleboard
- ☒ Needs no ironing
- ☐ Do not iron
- ☒ Fabric may shrink
- ☐ May bleed or fade
- ☒ Finish washes out

Where to find
- ☐ Any fabric store
- ☒ Major chain store
- ☒ Stores that carry high quality fabric
- ☐ Fabric club
- ☐ Mail order
- ☒ Wholesale supplier

A stiff debate

Many people prefer starch on their oxford shirts, but the popular stiffener may damage the cotton. Starch invades the porous fiber and wears away at it from the inside out. Telltale signs of starch damage include small holes that appear for no good reason, frayed collars and cuffs, and buttons that pop off. Starch also attracts silverfish to cotton, especially when the garment is stored. A starched garment will wear longer if it is washed at home occasionally to remove starch buildup.

oxford cloth

A soft, comfortable cotton shirting fabric made with a semi-basket weave. Usually, two thinner warp yarns are grouped and woven as one against thicker, softly spun filling yarns to form the modified basket weave. Variations are made by weaving two or four warps against two fillings. Better grades are made with combed cotton and mercerized to give the fabric a soft luster. Oxford cloth is usually white, but it may also be made with colored warps and white fillings or dyed in solid colors. Stripes are also common. Oxford cloth is one of four shirting fabrics originally made by a Scotch mill in the late 19th century. The others were named for Cambridge, Harvard and Yale.

How to use

Oxford cloth has an easy drape that falls into soft flares and folds. It may be gathered, shirred or pleated into a moderate fullness. The fabric wears well and launders easily, but it wrinkles badly and tends to shrink. Use to make fitted or semi-fitted shirts, skirts, dresses. Launder at home or send out.

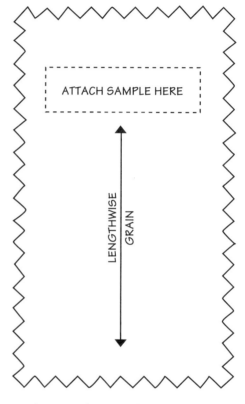

Cotton oxford is soft and smooth.

army oxford
A porous cotton shirting once used by the U.S. Army, but discontinued in the 1940s.

oxford chambray
A pastel oxford cloth woven like chambray, with colored warp yarns and white fillings. It is also called chambray oxford.

oxford stripes
Oxford cloth made with lengthwise colored stripes on a white background or white stripes on a colored ground. The stripes are made by alternating groups of white and colored warp yarns. When the filling yarns are white, the stripes are colored. Colored filling yarns produce white stripes.

pinpoint oxford
A superior grade of oxford cloth made in England and elsewhere with a tight weave and very fine combed cotton yarns. The popular shirting fabric is soft and silky. It is frequently white, off-white or pastel, with a soft luster from mercerization. It is used to make better quality men's dress shirts.

Sewing rating
- ☒ Easy to sew
- ☐ Moderately easy
- ☐ Average
- ☐ Moderately difficult
- ☐ Extremely difficult

Suggested fit
- ☐ Stretch to fit
- ☐ Close-fitting
- ☒ Fitted
- ☒ Semi-fitted
- ☒ Loose-fitting
- ☐ Very loose-fitting

Suggested styles
- ☒ Pleats ☐ Tucks
  - ☒ pressed
  - ☐ unpressed
- ☒ Gathers
  - ☐ limp ☒ soft
  - ☐ full ☐ lofty
  - ☐ bouffant
- ☒ Elasticized shirring
- ☐ Smocked
- ☒ Tailored
- ☐ Shaped with seams to eliminate bulk
- ☐ Lined
- ☐ Unlined
- ☐ Puffed or bouffant
- ☒ Loose and full
- ☒ Soft and flowing
- ☒ Draped
- ☐ Cut on bias
- ☐ Stretch styling

What to expect
- ☐ Difficult to cut out
- ☐ Fabric has one-way
  - ☐ design ☐ luster
  - ☐ weave ☐ nap
- ☒ Fabric is reversible
- ☒ It looks the same on both sides
- ☐ It stretches easily
- ☐ It will not stretch
- ☒ Fabric tears easily
- ☐ It is difficult to tear
- ☐ Fabric will not tear
- ☐ Pins and needles leave holes, marks
- ☐ It is difficult to ease sleeves and curves
- ☐ It tends to pucker
- ☐ It tends to unravel
- ☐ Inner construction shows from outside
- ☐ Machine eats fabric
- ☐ Skipped stitches
- ☐ Layers feed unevenly
- ☐ Multiple layers are difficult to cut, sew
- ☒ It creases easily
- ☐ Won't hold a crease

Cost per yard
- ☐ Less than $5
- ☒ $5 to $10
- ☒ $10 to $15
- ☐ $15 to $20
- ☐ $20 to $25
- ☐ More than $25

Semi-basket weave
Oxford cloth is made with a variation of the basket weave, called a semi-basket weave. In a basket weave, two or more yarns are grouped and woven as one yarn in both directions of the weave. In a semi-basket weave, the yarns are grouped in only one direction. For example, oxford cloth is made with lengthwise warp yarns that are half the size of the crosswise filling yarns, but the warp yarns are paired and woven as one yarn, producing a balanced weave with an unbalanced appearance. The semi-basket weave has a little give to it, but fabrics tend to wrinkle and shrink.

Wearability
- ☒ Durable ☐ Fragile
- ☐ Strong ☐ Weak
- ☒ It is long-wearing
- ☒ It wears evenly
- ☐ It wears out along seams and folds
- ☐ Seams don't hold up under stress
- ☐ Finish wears off
- ☐ Subject to abrasion
- ☐ It resists abrasion
- ☐ Subject to snags
- ☐ It resists snags
- ☐ Subject to runs
- ☐ It tends to pill
- ☐ It tends to shed
- ☐ It produces lint
- ☐ It attracts lint
- ☐ It attract static
- ☐ It tends to cling
- ☒ It holds its shape
- ☐ It loses its shape
- ☐ It stretches out of shape easily
- ☐ It droops, bags
- ☒ It tends to wrinkle
- ☐ It resists wrinkles
- ☐ It crushes easily
- ☐ Water drops leave spots or marks

Suggested care
- ☐ Dry clean only
- ☐ Do not dry clean
- ☒ Dry clean or wash
- ☐ Gently handwash in lukewarm water
- ☐ Roll in a towel to remove moisture
- ☐ Drip dry
- ☐ Lay flat to dry
- ☒ Machine wash
  - ☐ gentle/delicate
  - ☒ regular/normal
- ☒ Machine dry
  - ☐ cool ☐ normal
  - ☒ perm. press
- ☐ Press damp fabric
- ☒ Press dry fabric
- ☐ Dry iron ☒ Steam
- ☐ Iron on wrong side
- ☐ Use a press cloth
- ☐ Use a needleboard
- ☐ Needs no ironing
- ☐ Do not iron
- ☒ Fabric may shrink
- ☐ May bleed or fade
- ☒ Finish washes out

Where to find
- ☐ Any fabric store
- ☒ Major chain store
- ☒ Stores that carry high quality fabric
- ☐ Fabric club
- ☒ Mail order
- ☒ Wholesale supplier

Stuffer yarns

A stuffer yarn is an extra warp or filling yarn used to increase a fabric's weight, bulk, firmness or the prominence of its woven design. The stuffer yarn is not interwoven with the regular warp and filling yarns and it does not appear on the face of the fabric, although it may be visible on the back. The stuffer yarn is usually coarser and it may be made of cheaper material. It is easily removed from the fabric. The presence or absence of stuffer yarns is one way to determine the quality of fabrics like piqué.

piqué

A lightweight to heavyweight cotton fabric with a raised woven design. It usually has lengthwise cords or wales, but the pattern may also form another geometric shape. Better qualities are made of mercerized combed cotton with an extra set of warp yarns, called stuffer yarns, and a tight weave. The wales vary in size, depending on the size of the stuffer yarns. Poorer qualities are made from carded cotton without the stuffer yarns. Piqué may be bleached, printed or dyed, and preshrunk or slightly napped. It is sometimes made of rayon or silk.

How to use

Piqué has more body than flat fabrics. Lightweight versions may be gathered into a lofty fullness. Heavier fabrics work better when shaped with seams to eliminate bulk. Piqué is durable, but prominent floats on the back side usually wear out first. Fabric with prominent stuffer yarns is difficult to tear. Use to make fitted or semi-fitted blouses, dresses and summer suits. Piqué should be pressed on the wrong side to keep from flattening the raised design.

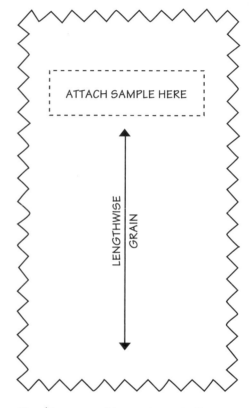

Piqué is woven with an extra set of yarns.

Bedford cord
A heavyweight fabric made the same way as piqué, with stuffer yarns inserted into the fabric to form lengthwise cords. It is usually cotton, but some versions may be wool or worsted with a cotton back. It is named for New Bedford, Massachusetts, and used to make trousers and uniforms.

bird's eye piqué
A fancy piqué with a tiny design formed by the wavy arrangement of the cords and by the use of stuffer yarns in both directions. Bull's eye pique has a similar, larger design.

pigskin piqué
A fancy piqué that resembles pigskin.

spiral piqué
A fancy piqué with a spiral design, made with stuffer yarns in both directions.

waffle piqué
A fine cotton fabric made with a small honeycomb weave. The fabric, which is fairly common, is not a true piqué.

Sewing rating
- ☐ Easy to sew
- ☐ Moderately easy
- ☒ Average
- ☐ Moderately difficult
- ☐ Extremely difficult

Suggested fit
- ☐ Stretch to fit
- ☒ Close-fitting
- ☒ Fitted
- ☒ Semi-fitted
- ☐ Loose-fitting
- ☐ Very loose-fitting

Suggested styles
- ☐ Pleats ☐ Tucks
  - ☐ pressed
  - ☐ unpressed
- ☒ Gathers
  - ☐ limp ☐ soft
  - ☐ full ☒ lofty
  - ☐ bouffant
- ☐ Elasticized shirring
- ☐ Smocked
- ☐ Tailored
- ☒ Shaped with seams to eliminate bulk
- ☒ Lined
- ☐ Unlined
- ☐ Puffed or bouffant
- ☐ Loose and full
- ☐ Soft and flowing
- ☐ Draped
- ☐ Cut on bias
- ☐ Stretch styling

What to expect
- ☐ Difficult to cut out
- ☒ Fabric has one-way
  - ☒ design ☐ luster
  - ☒ weave ☐ nap
- ☐ Fabric is reversible
- ☐ It looks the same on both sides
- ☐ It stretches easily
- ☐ It will not stretch
- ☐ Fabric tears easily
- ☒ It is difficult to tear
- ☒ Fabric will not tear
- ☐ Pins and needles leave holes, marks
- ☒ It is difficult to ease sleeves and curves
- ☐ It tends to pucker
- ☒ It tends to unravel
- ☐ Inner construction shows from outside
- ☐ Machine eats fabric
- ☐ Skipped stitches
- ☐ Layers feed unevenly
- ☒ Multiple layers are difficult to cut, sew
- ☐ It creases easily
- ☒ Won't hold a crease

Cost per yard
- ☐ Less than $5
- ☒ $5 to $10
- ☒ $10 to $15
- ☒ $15 to $20
- ☐ $20 to $25
- ☐ More than $25

The piqué weave
Piqué fabrics are made on a jacquard or dobby loom. Piquer is French for quilted, and indeed, the fabric's raised effect resembles quilting. The weave produces fabric with ridges, called wales or cords, formed by extra warp (lengthwise) yarns that are held in place on the back by crosswise filling floats. The extra yarns are called stuffer yarns because, like quilt stuffing, they do not show on the face of the fabric. The stuffer yarns are represented by black ovals in the illustration, surrounded by the regular warp yarns (small gray circles) and the filling yarns. Some versions have crosswise stuffer yarns or stuffer yarns in both directions.

Wearability
- ☒ Durable ☐ Fragile
- ☐ Strong ☐ Weak
- ☐ It is long-wearing
- ☐ It wears evenly
- ☒ It wears out along seams and folds
- ☐ Seams don't hold up under stress
- ☐ Finish wears off
- ☒ Subject to abrasion
- ☐ It resists abrasion
- ☒ Subject to snags
- ☐ It resists snags
- ☐ Subject to runs
- ☐ It tends to pill
- ☐ It tends to shed
- ☐ It produces lint
- ☐ It attracts lint
- ☐ It attract static
- ☐ It tends to cling
- ☒ It holds its shape
- ☐ It loses its shape
- ☐ It stretches out of shape easily
- ☐ It droops, bags
- ☐ It tends to wrinkle
- ☒ It resists wrinkles
- ☐ It crushes easily
- ☐ Water drops leave spots or marks

Suggested care
- ☐ Dry clean only
- ☐ Do not dry clean
- ☒ Dry clean or wash
- ☐ Gently handwash in lukewarm water
- ☐ Roll in a towel to remove moisture
- ☐ Drip dry
- ☐ Lay flat to dry
- ☒ Machine wash
  - ☒ gentle/delicate
  - ☐ regular/normal
- ☒ Machine dry
  - ☐ cool ☒ normal
  - ☐ perm. press
- ☐ Press damp fabric
- ☒ Press dry fabric
- ☐ Dry iron ☒ Steam
- ☒ Iron on wrong side
- ☐ Use a press cloth
- ☐ Use a needleboard
- ☐ Needs no ironing
- ☐ Do not iron
- ☒ Fabric may shrink
- ☐ May bleed or fade
- ☐ Finish washes out

Where to find
- ☐ Any fabric store
- ☒ Major chain store
- ☒ Stores that carry high quality fabric
- ☒ Fabric club
- ☒ Mail order
- ☒ Wholesale supplier

plissé

A lightweight cotton or cotton/polyester fabric printed with a caustic soda solution, usually in the form of stripes or random spots. The chemical shrinks parts of the fabric, causing the untreated parts to pucker. At first glance, plissé looks a lot like seersucker, but the two fabrics are different — seersucker's puckers are woven into the cloth. Plissé is flatter and not evenly puckered; some parts of the fabric may be obviously crinkled while other parts are almost flat. Plissé may be printed or dyed solid colors. The term is French for wrinkling, crinkling or pleating.

How to use

Plissé has a crisp springy drape that falls away from the body in crisp flares. It may be gathered, shirred or lightly tucked into a lofty fullness. The fabric is durable, but the puckers tend to flatten out in the wash. Use to make fitted, semi-fitted or loose-fitting casual clothing, children's clothing and sleepwear. Flattened puckers can be restored by the tumbling action of the dryer.

ATTACH SAMPLE HERE

LENGTHWISE GRAIN

Plissé has a crisp springy hand.

Embossed cotton

Embossed fabric is similar to plissé, but the raised design is produced by a different method. The pattern is pressed into the fabric with an engraved heated roller or plate. The results are permanent on synthetics, but cotton fabric must be treated with a resin finish to make the design more durable. Blends of cotton and polyester are more successful. Embossed fabrics require little or no ironing — the raised design may actually be pressed flat with too much steam, heat and pressure.

blister crêpe
A variation of caustic soda crêpe, printed with an all-over design instead of stripes. The process is usually applied to cotton fabrics such as lawn and organdy.

crinkle crêpe
Another name for plissé.

caustic soda crêpe
A crêpe effect produced by the application of caustic soda to part or all of a cotton fabric. Treated parts of the fabric shrink, causing untreated areas to pucker. The fabric is called plissé when the pattern forms stripes. The technique is also used to make other fabrics.

crêpe
A general classification for a wide variety of fabrics with grainy, pebbled or crinkled surfaces. True crêpe is woven with special tightly twisted crêpe yarns. Imitations are made with special weaves, embossed or treated with a chemical solution. The crêpe-like effect is not always permanent.

Sewing rating
- ☒ Easy to sew
- ☐ Moderately easy
- ☐ Average
- ☐ Moderately difficult
- ☐ Extremely difficult

Suggested fit
- ☐ Stretch to fit
- ☐ Close-fitting
- ☐ Fitted
- ☒ Semi-fitted
- ☒ Loose-fitting
- ☒ Very loose-fitting

Suggested styles
- ☐ Pleats ☒ Tucks
  - ☐ pressed
  - ☒ unpressed
- ☒ Gathers
  - ☐ limp ☐ soft
  - ☒ full ☒ lofty
  - ☐ bouffant
- ☒ Elasticized shirring
- ☐ Smocked
- ☐ Tailored
- ☐ Shaped with seams to eliminate bulk
- ☐ Lined
- ☐ Unlined
- ☒ Puffed or bouffant
- ☒ Loose and full
- ☐ Soft and flowing
- ☐ Draped
- ☐ Cut on bias
- ☐ Stretch styling

What to expect
- ☐ Difficult to cut out
- ☐ Fabric has one-way
  - ☐ design ☐ luster
  - ☐ weave ☐ nap
- ☐ Fabric is reversible
- ☐ It looks the same on both sides
- ☐ It stretches easily
- ☐ It will not stretch
- ☒ Fabric tears easily
- ☐ It is difficult to tear
- ☐ Fabric will not tear
- ☐ Pins and needles leave holes, marks
- ☐ It is difficult to ease sleeves and curves
- ☐ It tends to pucker
- ☐ It tends to unravel
- ☐ Inner construction shows from outside
- ☐ Machine eats fabric
- ☐ Skipped stitches
- ☐ Layers feed unevenly
- ☐ Multiple layers are difficult to cut, sew
- ☐ It creases easily
- ☒ Won't hold a crease

Cost per yard
- ☒ Less than $5
- ☒ $5 to $10
- ☐ $10 to $15
- ☐ $15 to $20
- ☐ $20 to $25
- ☐ More than $25

Caustic soda
Caustic soda is the common name for sodium hydroxide (NaOH), one of the most important alkaline compounds. A strong alkaline will dissolve animal fibers, such as wool, but is used in the production and manufacture of numerous textile products from cellulose fibers, including cotton, linen and viscose rayon. For example, to improve cotton's ability to absorb moisture, the fiber or fabric is boiled in a solution of caustic soda, which removes natural waxes and gums. This process is called **scouring** or **boiling off**. Sodium carbonate is another alkaline compound used on cotton. A strong acid will dissolve cellulose fibers, including cotton.

Wearability
- ☒ Durable ☐ Fragile
- ☐ Strong ☐ Weak
- ☒ It is long-wearing
- ☐ It wears evenly
- ☐ It wears out along seams and folds
- ☐ Seams don't hold up under stress
- ☐ Finish wears off
- ☐ Subject to abrasion
- ☐ It resists abrasion
- ☐ Subject to snags
- ☐ It resists snags
- ☐ Subject to runs
- ☐ It tends to pill
- ☐ It tends to shed
- ☐ It produces lint
- ☐ It attracts lint
- ☐ It attract static
- ☐ It tends to cling
- ☒ It holds its shape
- ☐ It loses its shape
- ☐ It stretches out of shape easily
- ☐ It droops, bags
- ☐ It tends to wrinkle
- ☒ It resists wrinkles
- ☐ It crushes easily
- ☐ Water drops leave spots or marks

Suggested care
- ☐ Dry clean only
- ☐ Do not dry clean
- ☐ Dry clean or wash
- ☐ Gently handwash in lukewarm water
- ☐ Roll in a towel to remove moisture
- ☐ Drip dry
- ☐ Lay flat to dry
- ☒ Machine wash
  - ☐ gentle/delicate
  - ☒ regular/normal
- ☒ Machine dry
  - ☐ cool ☒ normal
  - ☒ perm. press
- ☐ Press damp fabric
- ☐ Press dry fabric
- ☐ Dry iron ☐ Steam
- ☐ Iron on wrong side
- ☐ Use a press cloth
- ☐ Use a needleboard
- ☒ Needs no ironing
- ☒ Do not iron
- ☐ Fabric may shrink
- ☐ May bleed or fade
- ☒ Finish washes out

Where to find
- ☒ Any fabric store
- ☒ Major chain store
- ☐ Stores that carry high quality fabric
- ☐ Fabric club
- ☒ Mail order
- ☒ Wholesale supplier

85

poplin

A firm, durable, tightly woven fabric with fine crosswise ribs. On some fabrics, the ribs are formed by heavier filling yarns. On others, the fillings are the same size as the warps, but they are grouped to form the ribs. The dense fabric may have up to three times as many warp yarns as fillings per inch. The weight varies from light to medium. Poplin comes from *popeline*, French for a cloth made in the papal city of Avignon that was used for church vestmests. The original fabric had silk warp and wool filling. Today, it is usually made from a polyester/cotton blend, but it may also be all-cotton, synthetic or silk. Poplin is similar to broadcloth, but broadcloth is somewhat lighter in weight and has finer ribs.

How to use

Poplin has a moderately stiff drape that falls away from the body in wide flares. It may be be pleated into crisp folds or gathered into a lofty fullness. Polyester/cotton versions tend to pill. Use lighter weights to make shirts and dresses. Use heavier weights to make outer wear, especially rain coats.

Singin' in the rain

Poplin is popular for rain coats because the fabric is naturally water repellent. The tight weave works together with the fabric's crosswise ribs, which swell when they get wet, forming such a dense web that water from an ordinary rain shower will run off the surface rather than soak in. Polyester/cotton blends repel rain better than the all-cotton versions, because the synthetic fiber does not absorb water at all. The fabric is not waterproof, however. A good, hard rain will eventually soak through.

Poplin has fine crosswise ribs.

British poplin
A lightweight cotton shirting fabric with a fine crosswise rib, originally made in Great Britain. It was introduced to the United States in the 1920s, where the name was changed to broadcloth instead because the term "poplin" was thought to describe a heavier fabric.

Irish poplin
A fine fabric known for its uniform quality, made with silk organzine warp yarns and a worsted filling. The name suggests an Irish origin, but it was first made in China and later in Ireland. Today the fabric is not very common. The term also refers to a fine linen or cotton shirting made in Ireland.

popelin broché
A fine quality, brocaded or embossed silk poplin, made in France.

poplinette
British term for a very lightweight cotton poplin made with single yarns. It is usually called broadcloth in the United States.

Sewing rating
- [ ] Easy to sew
- [ ] Moderately easy
- [x] Average
- [ ] Moderately difficult
- [ ] Extremely difficult

Suggested fit
- [ ] Stretch to fit
- [x] Close-fitting
- [x] Fitted
- [x] Semi-fitted
- [x] Loose-fitting
- [ ] Very loose-fitting

Suggested styles
- [x] Pleats [ ] Tucks
  - [x] pressed
  - [ ] unpressed
- [x] Gathers
  - [ ] limp [ ] soft
  - [x] full [x] lofty
  - [ ] bouffant
- [x] Elasticized shirring
- [ ] Smocked
- [x] Tailored
- [ ] Shaped with seams to eliminate bulk
- [x] Lined
- [ ] Unlined
- [x] Puffed or bouffant
- [x] Loose and full
- [ ] Soft and flowing
- [ ] Draped
- [ ] Cut on bias
- [ ] Stretch styling

What to expect
- [ ] Difficult to cut out
- [ ] Fabric has one-way
  - [ ] design [ ] luster
  - [ ] weave [ ] nap
- [x] Fabric is reversible
- [x] It looks the same on both sides
- [ ] It stretches easily
- [ ] It will not stretch
- [x] Fabric tears easily
- [ ] It is difficult to tear
- [ ] Fabric will not tear
- [ ] Pins and needles leave holes, marks
- [x] It is difficult to ease sleeves and curves
- [ ] It tends to pucker
- [ ] It tends to unravel
- [ ] Inner construction shows from outside
- [ ] Machine eats fabric
- [ ] Skipped stitches
- [ ] Layers feed unevenly
- [ ] Multiple layers are difficult to cut, sew
- [x] It creases easily
- [ ] Won't hold a crease

Cost per yard
- [x] Less than $5
- [x] $5 to $10
- [ ] $10 to $15
- [ ] $15 to $20
- [ ] $20 to $25
- [ ] More than $25

The rib weave
Ribbed fabrics like poplin are made with the plain weave, but the warp-to-filling ratio is changed so that the yarns are more prominent in one direction than the other. The prominent yarns conceal the other yarns, forming ribs. The prominent yarns may be heavier, or several yarns may be woven together as one. When the warp yarns are heavier, lengthwise cords are formed. Crosswise ribs are produced by prominent filling yarns. The smaller yarns are not as strong and wear out faster. Fabrics with heavy ribs tend to unravel easily and slip at the seams.

Wearability
- [x] Durable [ ] Fragile
- [x] Strong [ ] Weak
- [x] It is long-wearing
- [x] It wears evenly
- [ ] It wears out along seams and folds
- [ ] Seams don't hold up under stress
- [ ] Finish wears off
- [ ] Subject to abrasion
- [ ] It resists abrasion
- [ ] Subject to snags
- [x] It resists snags
- [ ] Subject to runs
- [x] It tends to pill
- [ ] It tends to shed
- [ ] It produces lint
- [ ] It attracts lint
- [ ] It attract static
- [ ] It tends to cling
- [x] It holds its shape
- [ ] It loses its shape
- [ ] It stretches out of shape easily
- [ ] It droops, bags
- [ ] It tends to wrinkle
- [x] It resists wrinkles
- [ ] It crushes easily
- [ ] Water drops leave spots or marks

Suggested care
- [ ] Dry clean only
- [ ] Do not dry clean
- [x] Dry clean or wash
- [ ] Gently handwash in lukewarm water
- [ ] Roll in a towel to remove moisture
- [ ] Drip dry
- [ ] Lay flat to dry
- [x] Machine wash
  - [ ] gentle/delicate
  - [x] regular/normal
- [x] Machine dry
  - [ ] cool [ ] normal
  - [x] perm. press
- [ ] Press damp fabric
- [x] Press dry fabric
- [ ] Dry iron [x] Steam
- [ ] Iron on wrong side
- [ ] Use a press cloth
- [ ] Use a needleboard
- [ ] Needs no ironing
- [ ] Do not iron
- [x] Fabric may shrink
- [ ] May bleed or fade
- [x] Finish washes out

Where to find
- [x] Any fabric store
- [x] Major chain store
- [x] Stores that carry high quality fabric
- [x] Fabric club
- [x] Mail order
- [x] Wholesale supplier

[ATTACH SAMPLE HERE]

LENGTHWISE GRAIN

Sateen has a smooth, lustrous face.

sateen

Smooth cotton fabric with a soft lustrous face and a dull back, made with fine yarns, a high thread count and a satin weave. It has long floats that drift across the face of the fabric in a diagonal direction, giving it the appearance of a very flat twill. The best grades are woven with fine combed yarns and mercerized to increase the luster and smoothness. Colorful floral prints are common, but the fabric may also be dyed a solid color or woven with figured designs. Heavier fabrics are made with carded yarns and used as a base for coated fabrics or converted to ticking. The term originated as a way to distinguish cotton satin from a more typical satin made of lustrous fibers.

Natural luster

Cotton is not a very lustrous fiber, but it is often made more lustrous by treating the fabric with a special glaze or finish. A fabric also can be made more lustrous by the way it is woven. Sateen gets its luster from the smoothness of the weave, which reflects light better because the yarns float on the face. In other weaves, the yarns cross from front to back at frequent intervals, creating breaks on the face of the fabric similar to ripples on water. Sateen has fewer "ripples" on its face.

How to use

Sateen fabrics vary from soft and limp to somewhat crispy. Softer versions may be lightly gathered, while firmer fabrics work well with pleated styles. The tight weave is strong, but fabrics tend to snag and scuff. Use it for dresses, skirts and linings. Machine wash or dry clean to avoid damaging the face.

filing sateen

Cotton fabric made with a satin weave and a much greater number of crosswise filling yarns than lengthwise warp yarns. Filling yarns form floats on the face and warps cover the back. The filling yarns are softer and thicker than yarns used to make warp sateen, and fabric is not as fine.

sateen finish

A highly lustrous, crisp finish applied to some cotton fabrics to imitate sateen.

twill-faced filling sateen

A smooth, heavy, napped fabric made with carded yarns and a low-angle, right-hand twill, which produces a long filling float.

warp sateen

Also called cotton satin. Fabric is made with a satin weave and a greater number of lengthwise warp yarns than crosswise filling yarns. The warp yarns form floats on the face and the fillings cover the back. Warp yarns are finer, with a tighter twist, than yarns used to make filling sateen.

Sewing rating
- [ ] Easy to sew
- [ ] Moderately easy
- [x] Average
- [ ] Moderately difficult
- [ ] Extremely difficult

Suggested fit
- [ ] Stretch to fit
- [ ] Close-fitting
- [x] Fitted
- [x] Semi-fitted
- [x] Loose-fitting
- [ ] Very loose-fitting

Suggested styles
- [x] Pleats [ ] Tucks
  - [x] pressed
  - [ ] unpressed
- [x] Gathers
  - [ ] limp [ ] soft
  - [x] full [x] lofty
  - [ ] bouffant
- [x] Elasticized shirring
- [ ] Smocked
- [ ] Tailored
- [ ] Shaped with seams to eliminate bulk
- [ ] Lined
- [ ] Unlined
- [x] Puffed or bouffant
- [x] Loose and full
- [x] Soft and flowing
- [ ] Draped
- [ ] Cut on bias
- [ ] Stretch styling

What to expect
- [ ] Difficult to cut out
- [x] Fabric has one-way
  - [ ] design [x] luster
  - [ ] weave [ ] nap
- [ ] Fabric is reversible
- [ ] It looks the same on both sides
- [ ] It stretches easily
- [ ] It will not stretch
- [ ] Fabric tears easily
- [x] It is difficult to tear
- [ ] Fabric will not tear
- [ ] Pins and needles leave holes, marks
- [x] It is difficult to ease sleeves and curves
- [ ] It tends to pucker
- [ ] It tends to unravel
- [ ] Inner construction shows from outside
- [ ] Machine eats fabric
- [ ] Skipped stitches
- [ ] Layers feed unevenly
- [ ] Multiple layers are difficult to cut, sew
- [x] It creases easily
- [ ] Won't hold a crease

Cost per yard
- [ ] Less than $5
- [x] $5 to $10
- [x] $10 to $15
- [ ] $15 to $20
- [ ] $20 to $25
- [ ] More than $25

Satin weaves
Satin weaves are made for show, not durability. A typical satin weave has a greater number of yarns in one direction than the other, and those yarns, called "floats," are woven to appear mostly on the face of the fabric. Each face yarn "floats" over four or more yarns, under one and over four more, forming a smooth, lustrous surface. Satin has lengthwise warp floats, while sateen has crosswise filling floats. A filling-faced sateen cannot be woven as tightly as a warp-faced satin, so sateen fabrics tend to be coarser and heavier than true satins.

Wearability
- [ ] Durable [x] Fragile
- [x] Strong [ ] Weak
- [ ] It is long-wearing
- [ ] It wears evenly
- [x] It wears out along seams and folds
- [ ] Seams don't hold up under stress
- [x] Finish wears off
- [x] Subject to abrasion
- [ ] It resists abrasion
- [x] Subject to snags
- [ ] It resists snags
- [ ] Subject to runs
- [ ] It tends to pill
- [ ] It tends to shed
- [ ] It produces lint
- [ ] It attracts lint
- [ ] It attract static
- [ ] It tends to cling
- [ ] It holds its shape
- [ ] It loses its shape
- [ ] It stretches out of shape easily
- [ ] It droops, bags
- [x] It tends to wrinkle
- [ ] It resists wrinkles
- [ ] It crushes easily
- [x] Water drops leave spots or marks

Suggested care
- [ ] Dry clean only
- [ ] Do not dry clean
- [x] Dry clean or wash
- [ ] Gently handwash in lukewarm water
- [ ] Roll in a towel to remove moisture
- [ ] Drip dry
- [ ] Lay flat to dry
- [x] Machine wash
  - [x] gentle/delicate
  - [ ] regular/normal
- [x] Machine dry
  - [ ] cool [x] normal
  - [ ] perm. press
- [ ] Press damp fabric
- [x] Press dry fabric
- [ ] Dry iron [x] Steam
- [ ] Iron on wrong side
- [ ] Use a press cloth
- [ ] Use a needleboard
- [ ] Needs no ironing
- [x] Do not iron
- [x] Fabric may shrink
- [ ] May bleed or fade
- [x] Finish washes out

Where to find
- [ ] Any fabric store
- [x] Major chain store
- [x] Stores that carry high quality fabric
- [x] Fabric club
- [x] Mail order
- [x] Wholesale supplier

seersucker

A durable, firmly woven fabric with lengthwise puckered stripes alternating with flat ones. Seersucker is made by adjusting the yarn tension on the loom. Ordinary tension is applied to some groups of warp yarns and slack tension is applied to others. When the fabric is removed from the loom, the tighter yarns relax, causing the slack yarns to pucker permanently. The fabric is usually made of cotton or cotton and polyester, but it may also be rayon, silk or synthetic. It may be sheer or heavy, and solid in color or woven with dyed yarns to produce stripes, checks or plaids. Some versions are printed. Seersucker is not the same as plissé, which puckers because it is printed with a caustic soda.

How to use

Seersucker has a crisp drape that falls away from the body in wide flares. It may be pleated, gathered or shirred into a lofty fullness. Use to make fitted, semi-fitted or loose-fitting summer suits and casual clothing, sleepwear and children's clothing. The durable fabric launders well and needs no ironing.

Pucker up

The first seersucker was a linen cloth with alternating flat and puckered stripes, made in the East Indies. The fabric gets its name from the Persian word shirushakar, which translates literally to describe a cloth of "milk and sugar." In the 1930s, seersucker became popular in the United States for summer suits because the crisp, cool fabric did not show wrinkles and could be laundered easily. In the 1960s, wash and wear fabrics took its place, but it remains popular for other uses.

ATTACH SAMPLE HERE

LENGTHWISE GRAIN

Seersucker has flat and puckered stripes.

---

crinkle
Wrinkles or puckers in fabric produced by a number of methods, either during weaving of the cloth or in the finishing process.

crinkle cloth
A term used to describe any fabric with crinkles, including seersucker and plissé.

pucker
A puffed, crinkled, corrugated or blistered effect, produced by applying caustic soda or sulphuric acid to the fabric, by combining preshrunk and non-shrunk yarns, by using yarns that shrink differently, by printing the fabric with phenol paste, or by using different size yarns in a special weave.

puckered cloth
A blistered, crinkled or puffed cloth. It may be a desired effect or a flaw in the fabric.

seersucker gingham
A cotton fabric woven with dyed yarns in a checked pattern, like gingham, but with the puckered stripes of seersucker.

Sewing rating
- [ ] Easy to sew
- [ ] Moderately easy
- [x] Average
- [ ] Moderately difficult
- [ ] Extremely difficult

Suggested fit
- [ ] Stretch to fit
- [ ] Close-fitting
- [x] Fitted
- [x] Semi-fitted
- [x] Loose-fitting
- [ ] Very loose-fitting

Suggested styles
- [x] Pleats [x] Tucks
  - [ ] pressed
  - [x] unpressed
- [x] Gathers
  - [ ] limp [ ] soft
  - [ ] full [x] lofty
  - [ ] bouffant
- [x] Elasticized shirring
- [ ] Smocked
- [ ] Tailored
- [x] Shaped with seams to eliminate bulk
- [x] Lined
- [ ] Unlined
- [x] Puffed or bouffant
- [x] Loose and full
- [ ] Soft and flowing
- [ ] Draped
- [ ] Cut on bias
- [ ] Stretch styling

What to expect
- [ ] Difficult to cut out
- [ ] Fabric has one-way
  - [ ] design [ ] luster
  - [ ] weave [ ] nap
- [x] Fabric is reversible
- [x] It looks the same on both sides
- [ ] It stretches easily
- [ ] It will not stretch
- [ ] Fabric tears easily
- [x] It is difficult to tear
- [ ] Fabric will not tear
- [ ] Pins and needles leave holes, marks
- [x] It is difficult to ease sleeves and curves
- [ ] It tends to pucker
- [ ] It tends to unravel
- [ ] Inner construction shows from outside
- [ ] Machine eats fabric
- [ ] Skipped stitches
- [ ] Layers feed unevenly
- [ ] Multiple layers are difficult to cut, sew
- [ ] It creases easily
- [x] Won't hold a crease

Cost per yard
- [ ] Less than $5
- [x] $5 to $10
- [x] $10 to $15
- [ ] $15 to $20
- [ ] $20 to $25
- [ ] More than $25

Wrinkle busters
Some cotton fabrics don't wrinkle as easily as others. In general, plain weaves and twill fabrics tend to wrinkle more than satin, basket and rib weaves. A tightly woven fabric, like broadcloth, will wrinkle more than a loosely woven cloth, like gauze. Thin, fine yarns crease more easily than thick, heavy yarns. Usually, a fabric that doesn't wrinkle also won't hold a crease, so the garment style should be considered carefully. Other wrinkle busters include:
- Knits, such as double knit, interlock and jersey.
- Pile fabrics, such as corduroy and terry cloth.
- Rugged fabrics like canvas, denim, drill and duck.
- Fancy weaves, including dobby and piqué.

Wearability
- [x] Durable [ ] Fragile
- [x] Strong [ ] Weak
- [x] It is long-wearing
- [x] It wears evenly
- [ ] It wears out along seams and folds
- [ ] Seams don't hold up under stress
- [ ] Finish wears off
- [ ] Subject to abrasion
- [ ] It resists abrasion
- [ ] Subject to snags
- [ ] It resists snags
- [ ] Subject to runs
- [ ] It tends to pill
- [ ] It tends to shed
- [ ] It produces lint
- [ ] It attracts lint
- [ ] It attract static
- [ ] It tends to cling
- [x] It holds its shape
- [ ] It loses its shape
- [ ] It stretches out of shape easily
- [ ] It droops, bags
- [ ] It tends to wrinkle
- [x] It resists wrinkles
- [ ] It crushes easily
- [ ] Water drops leave spots or marks

Suggested care
- [ ] Dry clean only
- [ ] Do not dry clean
- [x] Dry clean or wash
- [ ] Gently handwash in lukewarm water
- [ ] Roll in a towel to remove moisture
- [ ] Drip dry
- [ ] Lay flat to dry
- [x] Machine wash
  - [ ] gentle/delicate
  - [x] regular/normal
- [x] Machine dry
  - [ ] cool [x] normal
  - [x] perm. press
- [ ] Press damp fabric
- [ ] Press dry fabric
- [ ] Dry iron [ ] Steam
- [ ] Iron on wrong side
- [ ] Use a press cloth
- [ ] Use a needleboard
- [x] Needs no ironing
- [x] Do not iron
- [ ] Fabric may shrink
- [ ] May bleed or fade
- [ ] Finish washes out

Where to find
- [ ] Any fabric store
- [x] Major chain store
- [x] Stores that carry high quality fabric
- [ ] Fabric club
- [x] Mail order
- [x] Wholesale supplier

shirting

General term for wide variety of lightweight fabrics used to make shirts. Fabrics of all fibers and blends are included, but it is most often used to describe cotton and cotton/polyester fabrics that do not fit the more specific descriptions of oxford cloth, broadcloth and chambray, among others. It may be made from coarse carded yarns or fine combed yarns in plain or fancy weaves, and it may be sized, mercerized, polished or treated to minimize wrinkles. Some of the best shirtings come from England. The most costly fabrics are made of extra-long staple Egyptian or Sea Island cotton.

How to use

Cotton shirtings have a soft or crisp drape, depending on the type of finish. Fabrics may be pleated, gathered or shirred into a moderate fullness, but tight weaves may be difficult to ease. Cotton/polyester versions may develop pills. Use to make fitted, semi-fitted or loose-fitting shirts and lightweight garments. Launder at home or dry clean for neat, sharply pressed results.

Cotton shirting with woven stripes.

> ### Durable press
> The terms "permanent press" and "durable press" are used to describe any garment or fabric that has been treated to retain its shape for life. Durable press is preferred because it is more accurate. The finishes reduce wrinkles and puckers. They also are used to improve flatness of seams, sharpness of creases, fabric smoothness and shape retention. Treated fabrics need no ironing, but they tend to be less comfortable because the finish reduces cotton's ability to breathe and absorb moisture.

Coneprest®
Registered trademark for durable-press fabrics made by Cone Mills Marketing Co. of cotton, synthetics and blends.

Dan-Press®
Registered trademark for a durable-press finish developed by Dan River, Inc.

Never-Press®
Registered trademark of Wamsutta Mills for a durable-press finish applied to all-cotton fabrics.

Permafresh®
Registered trademark for a durable-press finish developed by Sun Chemical Corp.

Reeve-Set®
Registered trademark for a durable-press finish developed by Reeves Brothers, Inc., of New York City.

Super-Crease®
Registered trademark for a durable-press finish developed by J. P. Stevens & Co., Inc.

Sewing rating
- [ ] Easy to sew
- [x] Moderately easy
- [ ] Average
- [ ] Moderately difficult
- [ ] Extremely difficult

Suggested fit
- [ ] Stretch to fit
- [ ] Close-fitting
- [x] Fitted
- [x] Semi-fitted
- [x] Loose-fitting
- [x] Very loose-fitting

Suggested styles
- [x] Pleats [x] Tucks
  - [x] pressed
  - [ ] unpressed
- [x] Gathers
  - [ ] limp [x] soft
  - [x] full [ ] lofty
  - [ ] bouffant
- [x] Elasticized shirring
- [ ] Smocked
- [x] Tailored
- [ ] Shaped with seams to eliminate bulk
- [ ] Lined
- [ ] Unlined
- [ ] Puffed or bouffant
- [x] Loose and full
- [x] Soft and flowing
- [ ] Draped
- [ ] Cut on bias
- [ ] Stretch styling

What to expect
- [ ] Difficult to cut out
- [ ] Fabric has one-way
  - [ ] design [ ] luster
  - [ ] weave [ ] nap
- [ ] Fabric is reversible
- [ ] It looks the same on both sides
- [ ] It stretches easily
- [ ] It will not stretch
- [x] Fabric tears easily
- [ ] It is difficult to tear
- [ ] Fabric will not tear
- [ ] Pins and needles leave holes, marks
- [x] It is difficult to ease sleeves and curves
- [ ] It tends to pucker
- [ ] It tends to unravel
- [ ] Inner construction shows from outside
- [ ] Machine eats fabric
- [ ] Skipped stitches
- [ ] Layers feed unevenly
- [ ] Multiple layers are difficult to cut, sew
- [x] It creases easily
- [ ] Won't hold a crease

Cost per yard
- [ ] Less than $5
- [x] $5 to $10
- [x] $10 to $15
- [ ] $15 to $20
- [ ] $20 to $25
- [x] More than $25

Sea Island cotton

Sea Island cotton is the temperamental aristocrat of cottons. The very white, lustrous fiber is strong, fine and remarkably uniform. The extra-long staple ranges from 1 3/4 to 2 1/4 inches long. It is usually combed, sometimes two or three times, before being spun into the finest of yarns and made into expensive fabrics. The cotton is named for the small islands off the coast of Georgia and the Carolinas, where it was raised beginning in the late 1700s. The fussy plant thrived in only a few areas near the sea and was hard to grow, but it made a few planters rich. Sea Island is the ancestor of today's finest hybrids, but it is rarely grown because of the cost.

Wearability
- [x] Durable [ ] Fragile
- [x] Strong [ ] Weak
- [x] It is long-wearing
- [x] It wears evenly
- [ ] It wears out along seams and folds
- [ ] Seams don't hold up under stress
- [ ] Finish wears off
- [ ] Subject to abrasion
- [ ] It resists abrasion
- [ ] Subject to snags
- [x] It resists snags
- [ ] Subject to runs
- [x] It tends to pill
- [ ] It tends to shed
- [ ] It produces lint
- [ ] It attracts lint
- [ ] It attract static
- [ ] It tends to cling
- [x] It holds its shape
- [ ] It loses its shape
- [ ] It stretches out of shape easily
- [ ] It droops, bags
- [x] It tends to wrinkle
- [ ] It resists wrinkles
- [ ] It crushes easily
- [ ] Water drops leave spots or marks

Suggested care
- [ ] Dry clean only
- [ ] Do not dry clean
- [x] Dry clean or wash
- [ ] Gently handwash in lukewarm water
- [ ] Roll in a towel to remove moisture
- [ ] Drip dry
- [ ] Lay flat to dry
- [x] Machine wash
  - [ ] gentle/delicate
  - [x] regular/normal
- [x] Machine dry
  - [ ] cool [ ] normal
  - [x] perm. press
- [ ] Press damp fabric
- [x] Press dry fabric
- [ ] Dry iron [x] Steam
- [ ] Iron on wrong side
- [ ] Use a press cloth
- [ ] Use a needleboard
- [ ] Needs no ironing
- [ ] Do not iron
- [x] Fabric may shrink
- [ ] May bleed or fade
- [x] Finish washes out

Where to find
- [ ] Any fabric store
- [x] Major chain store
- [x] Stores that carry high quality fabric
- [x] Fabric club
- [x] Mail order
- [x] Wholesale supplier

Absorbency

Most terry cloth is made with cotton because the absorbent fiber gets stronger when wet and it can be sanitized in very hot water, using strong bleach and detergent, without harm. Terry cloth also may be made with a cotton/polyester blend, which sounds like a bad idea at first because polyester does not absorb moisture. But the fabric usually has blended yarns only in the background weave and selvages to add strength and durability. The all-cotton loops absorb the moisture.

terry cloth

A plain or printed pile fabric with uncut loops on one or both sides, formed by inserting extra warp yarns into a firm, plain or twill background weave. The loops may cover the entire surface or form stripes, checks and other designs. Some versions have woven patterns made on a dobby or jacquard loom. Terry cloth has no nap. The weight varies, but it is usually a plush, soft and thick fabric. Cotton and rayon versions are the most absorbent, but it may also be made with a cotton/synthetic blend to increase the strength and durability.

How to use

Terry cloth has a moderately soft drape that falls into wide folds and flares. It may be gathered into a thick, heavy fullness or shaped with seams to reduce bulk. The fabric's thickness makes it difficult to cut and sew through multiple layers. Looped pile tends to snag and the fabric sheds lint readily, especially when new. Use to make bathrobes, beach wear, beach towels and casual garments. Terry cloth fluffs up in the dryer and needs no ironing.

ATTACH SAMPLE HERE

LENGTHWISE GRAIN

Terry cloth with uncut looped pile.

---

corduroy toweling
A terry cloth fabric with alternating looped stripes and plain stripes.

glass toweling
A smooth plain-weave fabric made with tightly twisted yarns to prevent shedding of loose fibers, used to make dish towels.

huck toweling
A soft, narrow cotton or linen fabric made with a loose open weave and a bird's eye or honeycomb pattern. It usually has strong selvages. Used to make kitchen towels.

knit terry
Knit fabric with a looped surface, like terry.

terry velvet
A velvet fabric made with uncut looped pile.

Turkish toweling
Thick, very absorbent terry cloth.

velour toweling
Terry cloth with sheared pile on one side.

Sewing rating
- [ ] Easy to sew
- [ ] Moderately easy
- [ ] Average
- [x] Moderately difficult
- [ ] Extremely difficult

Suggested fit
- [ ] Stretch to fit
- [ ] Close-fitting
- [x] Fitted
- [x] Semi-fitted
- [x] Loose-fitting
- [ ] Very loose-fitting

Suggested styles
- [ ] Pleats [ ] Tucks
  - [ ] pressed
  - [ ] unpressed
- [x] Gathers
  - [ ] limp [ ] soft
  - [x] full [ ] lofty
  - [ ] bouffant
- [ ] Elasticized shirring
- [ ] Smocked
- [ ] Tailored
- [x] Shaped with seams to eliminate bulk
- [ ] Lined
- [ ] Unlined
- [ ] Puffed or bouffant
- [x] Loose and full
- [x] Soft and flowing
- [ ] Draped
- [ ] Cut on bias
- [ ] Stretch styling

What to expect
- [x] Difficult to cut out
- [ ] Fabric has one-way
  - [ ] design [ ] luster
  - [ ] weave [ ] nap
- [x] Fabric is reversible
- [ ] It looks the same on both sides
- [ ] It stretches easily
- [x] It will not stretch
- [ ] Fabric tears easily
- [ ] It is difficult to tear
- [x] Fabric will not tear
- [ ] Pins and needles leave holes, marks
- [x] It is difficult to ease sleeves and curves
- [ ] It tends to pucker
- [ ] It tends to unravel
- [ ] Inner construction shows from outside
- [ ] Machine eats fabric
- [ ] Skipped stitches
- [ ] Layers feed unevenly
- [x] Multiple layers are difficult to cut, sew
- [ ] It creases easily
- [x] Won't hold a crease

Cost per yard
- [ ] Less than $5
- [x] $5 to $10
- [x] $10 to $15
- [ ] $15 to $20
- [ ] $20 to $25
- [ ] More than $25

Looped pile
Terry cloth is usually made with looped pile because the loops act like very small sponges, making the fabric more absorbent. Looped pile is also better able to withstand the strain of rubbing, pulling, twisting and tugging of the user and of constant laundering. The best terry cloth is made with a firm, tight twill background weave and closely packed loops. Loosely twisted loops are softer and more absorbent than tightly twisted loops, which produce a rougher fabric. Long pile is more absorbent than short pile. Terry cloth is most absorbent when it has loops on both sides.

Uncut, looped pile

Cut pile

Wearability
- [x] Durable [ ] Fragile
- [x] Strong [ ] Weak
- [x] It is long-wearing
- [x] It wears evenly
- [ ] It wears out along seams and folds
- [ ] Seams don't hold up under stress
- [ ] Finish wears off
- [ ] Subject to abrasion
- [ ] It resists abrasion
- [x] Subject to snags
- [ ] It resists snags
- [ ] Subject to runs
- [ ] It tends to pill
- [x] It tends to shed
- [x] It produces lint
- [ ] It attracts lint
- [ ] It attract static
- [ ] It tends to cling
- [x] It holds its shape
- [ ] It loses its shape
- [ ] It stretches out of shape easily
- [ ] It droops, bags
- [ ] It tends to wrinkle
- [x] It resists wrinkles
- [ ] It crushes easily
- [ ] Water drops leave spots or marks

Suggested care
- [ ] Dry clean only
- [ ] Do not dry clean
- [ ] Dry clean or wash
- [ ] Gently handwash in lukewarm water
- [ ] Roll in a towel to remove moisture
- [ ] Drip dry
- [ ] Lay flat to dry
- [x] Machine wash
  - [ ] gentle/delicate
  - [x] regular/normal
- [x] Machine dry
  - [ ] cool [x] normal
  - [ ] perm. press
- [ ] Press damp fabric
- [ ] Press dry fabric
- [ ] Dry iron [ ] Steam
- [ ] Iron on wrong side
- [ ] Use a press cloth
- [ ] Use a needleboard
- [x] Needs no ironing
- [ ] Do not iron
- [ ] Fabric may shrink
- [x] May bleed or fade
- [ ] Finish washes out

Where to find
- [x] Any fabric store
- [x] Major chain store
- [x] Stores that carry high quality fabric
- [x] Fabric club
- [x] Mail order
- [x] Wholesale supplier

[ATTACH SAMPLE HERE]

LENGTHWISE GRAIN

Ticking stripes are usually blue and white.

ticking

A large group of firm, durable, tightly woven cotton and linen fabrics used to cover box springs, pillows and mattresses. Ticking may be made with a plain, twill, jacquard, herringbone or satin weave. Good quality ticking has more warp yarns than filling yarns per inch to add strength in the lengthwise direction. The weight varies from medium to heavy. Ticking usually has woven blue and white stripes, but it may also be plain or printed with large floral designs. Some versions are mercerized or schreinered to increase the fabric's strength and/or luster. Ticking comes from the Latin *theca*, meaning a cover or case made of cotton or sometimes of linen.

How to use

Ticking has a stiff drape that falls into crisp folds and wide cones. The fabric works best in garments that are shaped by seams to eliminate bulk. Use to make close-fitting or fitted pants and casual clothing, or to cover pillows and mattresses. Large bulky items may need to be dry cleaned.

Stuffing

Some stuffing comes from the seed pods of the Kapok tree, which grows in the tropics all over the world, but especially around Java. The lustrous fiber is sometimes called silk cotton or Java cotton, but it is neither silk nor cotton. Kapok fiber is cellulose, like cotton, but it is too brittle to spin into yarn. Kapok is resistant to moisture, naturally buoyant and resilient, making it ideal for stuffing life preservers, mattresses and pillows. It also is used to make insulation.

A.C.A. ticking

A strong, closely woven, warp-faced cotton twill with traditional ticking stripes — dark blue and natural or white. The term was once a trademark of Amoskeag Company for its A-quality ticking, but it is now used generically to describe better grades of ticking made by a number of textile mills.

Bohemian ticking

A featherproof pillow ticking made with a tight plain weave and given a calendered, waxed finish. It may be iridescent — woven with colored warp and white filling yarns. It may also be bleached white or have narrow colored stripes on white. Better grades are imported from Holland and elsewhere.

damask ticking

Mercerized cotton or cotton/rayon fabric with elaborate designs made on a jacquard loom, using contrasting colors in the warp and filling. Cotton/rayon versions are quite lustrous, but the all-cotton fabric is more durable. Damask ticking is expensive and is used to cover better quality mattresses.

Sewing rating
- [ ] Easy to sew
- [ ] Moderately easy
- [x] Average
- [ ] Moderately difficult
- [ ] Extremely difficult

Suggested fit
- [ ] Stretch to fit
- [x] Close-fitting
- [x] Fitted
- [x] Semi-fitted
- [ ] Loose-fitting
- [ ] Very loose-fitting

Suggested styles
- [ ] Pleats [ ] Tucks
  - [ ] pressed
  - [ ] unpressed
- [ ] Gathers
  - [ ] limp [ ] soft
  - [ ] full [ ] lofty
  - [ ] bouffant
- [ ] Elasticized shirring
- [ ] Smocked
- [ ] Tailored
- [x] Shaped with seams to eliminate bulk
- [ ] Lined
- [ ] Unlined
- [ ] Puffed or bouffant
- [ ] Loose and full
- [ ] Soft and flowing
- [ ] Draped
- [ ] Cut on bias
- [ ] Stretch styling

What to expect
- [ ] Difficult to cut out
- [ ] Fabric has one-way
  - [ ] design [ ] luster
  - [ ] weave [ ] nap
- [ ] Fabric is reversible
- [ ] It looks the same on both sides
- [ ] It stretches easily
- [x] It will not stretch
- [ ] Fabric tears easily
- [x] It is difficult to tear
- [x] Fabric will not tear
- [ ] Pins and needles leave holes, marks
- [x] It is difficult to ease sleeves and curves
- [ ] It tends to pucker
- [ ] It tends to unravel
- [ ] Inner construction shows from outside
- [ ] Machine eats fabric
- [ ] Skipped stitches
- [ ] Layers feed unevenly
- [x] Multiple layers are difficult to cut, sew
- [ ] It creases easily
- [x] Won't hold a crease

Cost per yard
- [ ] Less than $5
- [x] $5 to $10
- [ ] $10 to $15
- [ ] $15 to $20
- [ ] $20 to $25
- [ ] More than $25

Featherproof ticking
The best pillows, comforters and beds are stuffed with the feathers and down of ducks, geese and other birds. The stuffing's natural loft provides comfort and warmth without a lot of weight. But feathers also have a less-desirable trait – sharp, pointed quills that poke holes into fabric. The best tickings are featherproof, made with a tight weave, strong warp yarns and softly spun filling yarns, which spread to keep the stuffing from poking its way out through the fabric. Usually, featherproof ticking is woven with many more warp yarns than filling yarns per inch.

Wearability
- [x] Durable [ ] Fragile
- [x] Strong [ ] Weak
- [x] It is long-wearing
- [x] It wears evenly
- [ ] It wears out along seams and folds
- [ ] Seams don't hold up under stress
- [ ] Finish wears off
- [ ] Subject to abrasion
- [x] It resists abrasion
- [ ] Subject to snags
- [x] It resists snags
- [ ] Subject to runs
- [ ] It tends to pill
- [ ] It tends to shed
- [ ] It produces lint
- [ ] It attracts lint
- [ ] It attract static
- [ ] It tends to cling
- [x] It holds its shape
- [ ] It loses its shape
- [ ] It stretches out of shape easily
- [ ] It droops, bags
- [ ] It tends to wrinkle
- [x] It resists wrinkles
- [ ] It crushes easily
- [x] Water drops leave spots or marks

Suggested care
- [ ] Dry clean only
- [ ] Do not dry clean
- [x] Dry clean or wash
- [ ] Gently handwash in lukewarm water
- [ ] Roll in a towel to remove moisture
- [ ] Drip dry
- [ ] Lay flat to dry
- [x] Machine wash
  - [ ] gentle/delicate
  - [x] regular/normal
- [x] Machine dry
  - [ ] cool [x] normal
  - [ ] perm. press
- [ ] Press damp fabric
- [ ] Press dry fabric
- [ ] Dry iron [ ] Steam
- [ ] Iron on wrong side
- [ ] Use a press cloth
- [ ] Use a needleboard
- [x] Needs no ironing
- [ ] Do not iron
- [x] Fabric may shrink
- [ ] May bleed or fade
- [x] Finish washes out

Where to find
- [ ] Any fabric store
- [x] Major chain store
- [x] Stores that carry high quality fabric
- [ ] Fabric club
- [x] Mail order
- [x] Wholesale supplier

Herringbone

Herringbone, a woven pattern that resembles the skeleton of a herring, is a popular variation of twill. It is made with a broken twill pattern that reverses the direction of the diagonal line at regular intervals. The fabric has alternating lengthwise stripes of left- and right-hand twill lines that form a chevron pattern when paired. The pattern is more obvious when woven with two colors of yarn. Solid colors are subtle and difficult to see. The herringbone weave is used to make all kinds of fabrics.

twill

General term for a variety of cotton fabrics made with the twill weave. There are many types, including denim, gabardine, ticking, chino and drill. A fabric that does not fit one of those descriptions is often called twill without any further distinction. The twill weave produces a durable fabric that does not rip or tear easily. A very steep twill line is a sign of better quality and greater strength. Such fabrics have more warp yarns per inch. Most cotton twills are medium to heavy in weight. Fabrics may be dyed, printed or bleached. Some are mercerized and treated to reduce wrinkles and shrinkage.

How to use

Generally speaking, cotton twill has a moderately stiff drape that holds the shape of the garment. Fabrics work best with styles that are shaped with seams to eliminate bulk. Tight weaves are difficult to ease. Use to make close-fitting, fitted or semi-fitted uniforms and casual slacks. Machine wash and tumble dry, but beware of shrinkage. Ironing is usually not required.

ATTACH SAMPLE HERE

LENGTHWISE GRAIN

This twill has a prominent diagonal line.

---

fancy twill
A general term for any twill fabric made with a combination of two or more regular twill weaves to produce a pattern different than the usual diagonal twill line.

fine-line twill
Used to describe a very fine diagonal line.

gabardine
A fine, tightly woven fabric with a distinct twill line on the right side and a plain back, usually made of wool. Cotton gabardine is elegant, strong and compact, but it is not very common. Lighter versions are used for designer sportswear, dresses and shirts, heavier fabrics for casual slacks.

whipcord
A compact twill fabric similar to gabardine with a very steep, prominent twill line on the right side and a plain back. The twill line is steeper than gabardine's. Woolen and worsted versions are used to make suits and topcoats, while cotton versions are used to make uniforms and casual slacks.

Sewing rating
- ☐ Easy to sew
- ☐ Moderately easy
- ☒ Average
- ☐ Moderately difficult
- ☐ Extremely difficult

Suggested fit
- ☐ Stretch to fit
- ☒ Close-fitting
- ☒ Fitted
- ☒ Semi-fitted
- ☐ Loose-fitting
- ☐ Very loose-fitting

Suggested styles
- ☐ Pleats ☐ Tucks
  - ☐ pressed
  - ☐ unpressed
- ☐ Gathers
  - ☐ limp ☐ soft
  - ☐ full ☐ lofty
  - ☐ bouffant
- ☐ Elasticized shirring
- ☐ Smocked
- ☐ Tailored
- ☒ Shaped with seams to eliminate bulk
- ☐ Lined
- ☐ Unlined
- ☐ Puffed or bouffant
- ☐ Loose and full
- ☐ Soft and flowing
- ☐ Draped
- ☐ Cut on bias
- ☐ Stretch styling

What to expect
- ☐ Difficult to cut out
- ☒ Fabric has one-way
  - ☐ design ☐ luster
  - ☒ weave ☐ nap
- ☐ Fabric is reversible
- ☐ It looks the same on both sides
- ☐ It stretches easily
- ☐ It will not stretch
- ☐ Fabric tears easily
- ☒ It is difficult to tear
- ☐ Fabric will not tear
- ☐ Pins and needles leave holes, marks
- ☒ It is difficult to ease sleeves and curves
- ☐ It tends to pucker
- ☐ It tends to unravel
- ☐ Inner construction shows from outside
- ☐ Machine eats fabric
- ☐ Skipped stitches
- ☐ Layers feed unevenly
- ☒ Multiple layers are difficult to cut, sew
- ☐ It creases easily
- ☒ Won't hold a crease

Cost per yard
- ☒ Less than $5
- ☒ $5 to $10
- ☒ $10 to $15
- ☐ $15 to $20
- ☐ $20 to $25
- ☐ More than $25

Broken twills
Many twill fabrics are made with a diagonal line that changes direction, forming a pattern that looks like a series of points, zigzags or chevrons. The pattern, called a broken twill, may be even, like the precise, V-shaped chevrons of herringbone, or uneven, with the twill line making more random changes in direction. On some broken twills, the diagonal line disappears. Houndstooth, for example, is made with a broken twill weave that forms a sort of ragged, checked pattern. There are many variations of broken twill, which is also called fancy, zigzag and pointed twill.

Wearability
- ☒ Durable ☐ Fragile
- ☒ Strong ☐ Weak
- ☒ It is long-wearing
- ☐ It wears evenly
- ☒ It wears out along seams and folds
- ☐ Seams don't hold up under stress
- ☐ Finish wears off
- ☐ Subject to abrasion
- ☐ It resists abrasion
- ☐ Subject to snags
- ☐ It resists snags
- ☐ Subject to runs
- ☐ It tends to pill
- ☐ It tends to shed
- ☐ It produces lint
- ☐ It attracts lint
- ☐ It attract static
- ☐ It tends to cling
- ☒ It holds its shape
- ☐ It loses its shape
- ☐ It stretches out of shape easily
- ☐ It droops, bags
- ☐ It tends to wrinkle
- ☒ It resists wrinkles
- ☐ It crushes easily
- ☐ Water drops leave spots or marks

Suggested care
- ☐ Dry clean only
- ☐ Do not dry clean
- ☐ Dry clean or wash
- ☐ Gently handwash in lukewarm water
- ☐ Roll in a towel to remove moisture
- ☐ Drip dry
- ☐ Lay flat to dry
- ☒ Machine wash
  - ☐ gentle/delicate
  - ☒ regular/normal
- ☒ Machine dry
  - ☐ cool ☒ normal
  - ☐ perm. press
- ☐ Press damp fabric
- ☒ Press dry fabric
- ☐ Dry iron ☒ Steam
- ☐ Iron on wrong side
- ☐ Use a press cloth
- ☐ Use a needleboard
- ☐ Needs no ironing
- ☐ Do not iron
- ☒ Fabric may shrink
- ☐ May bleed or fade
- ☒ Finish washes out

Where to find
- ☒ Any fabric store
- ☒ Major chain store
- ☒ Stores that carry high quality fabric
- ☒ Fabric club
- ☒ Mail order
- ☒ Wholesale supplier

velveteen

Soft, smooth fabric with short cropped pile, made with a plain or twill weave and an extra set of softly spun filling yarns, which are cut to form the pile. Better qualities are made with combed, ply warp yarns, while poorer qualities have single warp yarns. The extra filling is usually of long-staple, mercerized combed cotton. The density of the pile is determined by the number of filling yarns per inch, which may be as few as 175 or as many as 600, with about 80 warp yarns per inch. Pile may be cut by hand or machine. The weight varies, but velveteen is usually a thick, heavy fabric. Most versions are dyed a solid color, but it may be printed. A wax finish is sometimes applied to the pile to increase the luster.

How to use

Velveteen has a moderately soft drape that falls into wide folds and flares. It may be gathered into a lofty fullness or shaped with seams to eliminate bulk. The popular fabric is used for dressy and casual garments, bedspreads and draperies. Dry clean or machine wash and tumble dry, but remove promptly.

Velveteen has short, closely cropped pile.

Velvet

Velveteen is often considered to be the cotton version of velvet, but fiber content is only one of the differences between the two fabrics. Velvet is made with an extra set of warp yarns, while velveteen has extra filling yarns. Velvet has a richer, deeper, more erect pile. Velveteen's pile slopes slightly, given the fabric a flat, closely cropped appearance. The first velvet was silk but today it is more often made of rayon or other fibers. The plush fabric is softer, more lustrous and more luxurious than velveteen.

---

cotton velvet
A soft, plush fabric made the same way as other velvets, with an extra set of warp yarns to form the pile. The fabric is not as flat as velveteen and the pile is more erect. It made its debut during World War II when the source of silk was cut off, forcing mills to experiment with other fibers.

velour
This term, which is French for velvet, is loosely applied to a wide range of cut pile fabrics, but it most often refers to woven or knitted cotton fabric with a soft, dense closely cropped pile. It was originally made of wool, but is now made from other fibers. The knitted version is frequently synthetic.

velveteen plush
A soft, plush cotton fabric that is very similar to velveteen, but it is woven with longer filling floats, producing a pile that is longer and higher than the pile of velvet or regular velveteen. The pile of velveteen is closely cropped and is generally not more than 1/16 of an inch in length.

Sewing rating
- ☐ Easy to sew
- ☐ Moderately easy
- ☒ Average
- ☐ Moderately difficult
- ☐ Extremely difficult

Suggested fit
- ☐ Stretch to fit
- ☒ Close-fitting
- ☒ Fitted
- ☒ Semi-fitted
- ☐ Loose-fitting
- ☐ Very loose-fitting

Suggested styles
- ☐ Pleats ☐ Tucks
  - ☐ pressed
  - ☒ unpressed
- ☒ Gathers
  - ☐ limp ☐ soft
  - ☒ full ☒ lofty
  - ☐ bouffant
- ☐ Elasticized shirring
- ☐ Smocked
- ☒ Tailored
- ☒ Shaped with seams to eliminate bulk
- ☒ Lined
- ☐ Unlined
- ☒ Puffed or bouffant
- ☒ Loose and full
- ☐ Soft and flowing
- ☐ Draped
- ☐ Cut on bias
- ☐ Stretch styling

What to expect
- ☐ Difficult to cut out
- ☒ Fabric has one-way
  - ☐ design ☒ luster
  - ☐ weave ☒ nap
- ☐ Fabric is reversible
- ☐ It looks the same on both sides
- ☐ It stretches easily
- ☐ It will not stretch
- ☐ Fabric tears easily
- ☒ It is difficult to tear
- ☒ Fabric will not tear
- ☒ Pins and needles leave holes, marks
- ☒ It is difficult to ease sleeves and curves
- ☐ It tends to pucker
- ☐ It tends to unravel
- ☐ Inner construction shows from outside
- ☐ Machine eats fabric
- ☐ Skipped stitches
- ☒ Layers feed unevenly
- ☒ Multiple layers are difficult to cut, sew
- ☐ It creases easily
- ☒ Won't hold a crease

Cost per yard
- ☐ Less than $5
- ☐ $5 to $10
- ☒ $10 to $15
- ☒ $15 to $20
- ☒ $20 to $25
- ☐ More than $25

Take it to the cleaners
Cotton is easily laundered at home, but some fabrics, including velveteen, may be dry cleaned instead. Watch out for:
- ✓ Embossed designs.
- ✓ Fabrics or garments that are difficult to iron, such as evening wear, wedding gowns and large table cloths.
- ✓ Loose weaves and long floats that snag easily.
- ✓ Fabrics that tend to shrink or unravel.
- ✓ Delicate embellished fabrics, like dotted Swiss.
- ✓ Linings, shoulder pads and inner construction.
- ✓ Special finishes that come out in the wash.
- ✓ Garments that require professional pressing and finishing, such as starched men's shirts.

SIZE 8 — 100% COTTON MADE IN U.S.A. — DRY CLEAN ONLY

Wearability
- ☒ Durable ☐ Fragile
- ☒ Strong ☐ Weak
- ☒ It is long-wearing
- ☐ It wears evenly
- ☐ It wears out along seams and folds
- ☐ Seams don't hold up under stress
- ☒ Finish wears off
- ☒ Subject to abrasion
- ☐ It resists abrasion
- ☐ Subject to snags
- ☐ It resists snags
- ☐ Subject to runs
- ☐ It tends to pill
- ☒ It tends to shed
- ☒ It produces lint
- ☐ It attracts lint
- ☐ It attract static
- ☐ It tends to cling
- ☒ It holds its shape
- ☐ It loses its shape
- ☐ It stretches out of shape easily
- ☐ It droops, bags
- ☐ It tends to wrinkle
- ☒ It resists wrinkles
- ☐ It crushes easily
- ☒ Water drops leave spots or marks

Suggested care
- ☐ Dry clean only
- ☐ Do not dry clean
- ☒ Dry clean or wash
- ☐ Gently handwash in lukewarm water
- ☐ Roll in a towel to remove moisture
- ☐ Drip dry
- ☐ Lay flat to dry
- ☒ Machine wash
  - ☐ gentle/delicate
  - ☒ regular/normal
- ☒ Machine dry
  - ☐ cool ☐ normal
  - ☒ perm. press
- ☐ Press damp fabric
- ☒ Press dry fabric
- ☐ Dry iron ☒ Steam
- ☒ Iron on wrong side
- ☐ Use a press cloth
- ☒ Use a needleboard
- ☐ Needs no ironing
- ☐ Do not iron
- ☒ Fabric may shrink
- ☐ May bleed or fade
- ☒ Finish washes out

Where to find
- ☐ Any fabric store
- ☐ Major chain store
- ☒ Stores that carry high quality fabric
- ☐ Fabric club
- ☒ Mail order
- ☒ Wholesale supplier

Voile twist

Voile is made with special voile yarns, which are twisted more than average yarns but less than crêpe yarns. A voile twist (30-40 turns per inch) is also called a hard twist, because the twist produces a compact, firm yarn by forcing the fibers closer together. A two-ply voile yarn, or twist-on-twist yarn, is made by twisting each single yarn, then twisting the two plys together in the same direction as the individual twist. This produces a more pronounced "hardness" than single yarns.

voile

A semi-sheer, dainty fabric made with tightly twisted voile yarns and a loose plain weave, usually with the same number of yarns in both directions. Better grades are made with two-ply yarns of combed cotton. The fabric is not as tightly woven as lawn or as soft as batiste. It may be mercerized or sized to give it a crisp, wiry hand. Voile is usually made of cotton, polyester or a cotton/polyester blend. It is sometimes made of rayon, acetate, worsted or silk. The fabric may be a solid color, printed, striped or woven with novelty yarns for special effects. Voile is French for veil and was originally used to make veils.

How to use

Voile has a delicate drape that falls into graceful folds. It may be gathered or shirred into a soft or moderately crisp fullness. The semi-sheer fabric is popular for curtains because it provides privacy without blocking the light. It may also be used to make semi-fitted, loose or very loose blouses, dresses, lingerie and children's clothing. Machine wash and tumble dry on gentle cycles.

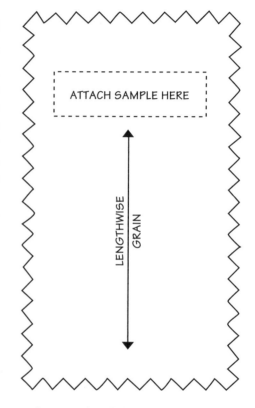

Cotton voile is lightweight and sheer.

---

dimity
A wide variety of sheer, lightweight cotton fabrics with fine lengthwise cords, made by weaving groups of two, three or more warp yarns together. A checked effect is made by grouping yarns in both directions, using a fancy basket weave. Dimity comes from dimitos, a Greek word for "double thread."

piqué voile
Voile made with lengthwise yarns spaced to produce a corded effect similar to piqué. Also called piqué dimity and corded voile.

seed voile
Cotton voile with a seed effect, made with slubbed novelty yarns called seed yarns.

shadow-stripe voile
A striped voile made by arranging the warp yarns closer together in the stripes than in the rest of the fabric.

splash voile
Cotton voile with occasional slubs in either direction, made with a special type of yarn.

Sewing rating
- [ ] Easy to sew
- [x] Moderately easy
- [ ] Average
- [ ] Moderately difficult
- [ ] Extremely difficult

Suggested fit
- [ ] Stretch to fit
- [ ] Close-fitting
- [ ] Fitted
- [x] Semi-fitted
- [x] Loose-fitting
- [x] Very loose-fitting

Suggested styles
- [ ] Pleats [ ] Tucks
  - [ ] pressed
  - [ ] unpressed
- [x] Gathers
  - [x] limp [x] soft
  - [ ] full [ ] lofty
  - [ ] bouffant
- [x] Elasticized shirring
- [ ] Smocked
- [ ] Tailored
- [ ] Shaped with seams to eliminate bulk
- [ ] Lined
- [ ] Unlined
- [ ] Puffed or bouffant
- [x] Loose and full
- [x] Soft and flowing
- [x] Draped
- [ ] Cut on bias
- [ ] Stretch styling

What to expect
- [ ] Difficult to cut out
- [ ] Fabric has one-way
  - [ ] design [ ] luster
  - [ ] weave [ ] nap
- [ ] Fabric is reversible
- [ ] It looks the same on both sides
- [ ] It stretches easily
- [ ] It will not stretch
- [x] Fabric tears easily
- [ ] It is difficult to tear
- [ ] Fabric will not tear
- [x] Pins and needles leave holes, marks
- [ ] It is difficult to ease sleeves and curves
- [ ] It tends to pucker
- [ ] It tends to unravel
- [x] Inner construction shows from outside
- [x] Machine eats fabric
- [ ] Skipped stitches
- [ ] Layers feed unevenly
- [ ] Multiple layers are difficult to cut, sew
- [x] It creases easily
- [ ] Won't hold a crease

Cost per yard
- [ ] Less than $5
- [x] $5 to $10
- [x] $10 to $15
- [x] $15 to $20
- [ ] $20 to $25
- [ ] More than $25

Yarn sizes
Sheer, light fabrics like voile and lawn are made with very fine threads, called yarns by the textile industry. Cotton yarns are numbered according to size. The numbering system is based on the number of hanks (840 yards) in 1 pound of yarn. The bigger the number, the finer the yarn. A standard yarn size is about 50, a very fine yarn is 80 to 100 and a coarse yarn is about 20. Single yarns also are marked with the letter s, such as 80s. Ply yarns carry a second number, such as 80/2, meaning the 40-weight yarn is made with two 80s. Voile is typically made with 70s to 100s, while 13s and 20s are used to make heavy denims and canvas.

Wearability
- [ ] Durable [x] Fragile
- [ ] Strong [ ] Weak
- [ ] It is long-wearing
- [x] It wears evenly
- [ ] It wears out along seams and folds
- [ ] Seams don't hold up under stress
- [x] Finish wears off
- [x] Subject to abrasion
- [ ] It resists abrasion
- [ ] Subject to snags
- [ ] It resists snags
- [ ] Subject to runs
- [ ] It tends to pill
- [ ] It tends to shed
- [ ] It produces lint
- [ ] It attracts lint
- [ ] It attract static
- [ ] It tends to cling
- [x] It holds its shape
- [ ] It loses its shape
- [ ] It stretches out of shape easily
- [ ] It droops, bags
- [x] It tends to wrinkle
- [ ] It resists wrinkles
- [x] It crushes easily
- [x] Water drops leave spots or marks

Suggested care
- [ ] Dry clean only
- [ ] Do not dry clean
- [ ] Dry clean or wash
- [ ] Gently handwash in lukewarm water
- [ ] Roll in a towel to remove moisture
- [ ] Drip dry
- [ ] Lay flat to dry
- [x] Machine wash
  - [x] gentle/delicate
  - [ ] regular/normal
- [x] Machine dry
  - [x] cool [ ] normal
  - [ ] perm. press
- [ ] Press damp fabric
- [x] Press dry fabric
- [ ] Dry iron [x] Steam
- [ ] Iron on wrong side
- [ ] Use a press cloth
- [ ] Use a needleboard
- [ ] Needs no ironing
- [ ] Do not iron
- [x] Fabric may shrink
- [x] May bleed or fade
- [x] Finish washes out

Where to find
- [ ] Any fabric store
- [ ] Major chain store
- [x] Stores that carry high quality fabric
- [ ] Fabric club
- [x] Mail order
- [x] Wholesale supplier

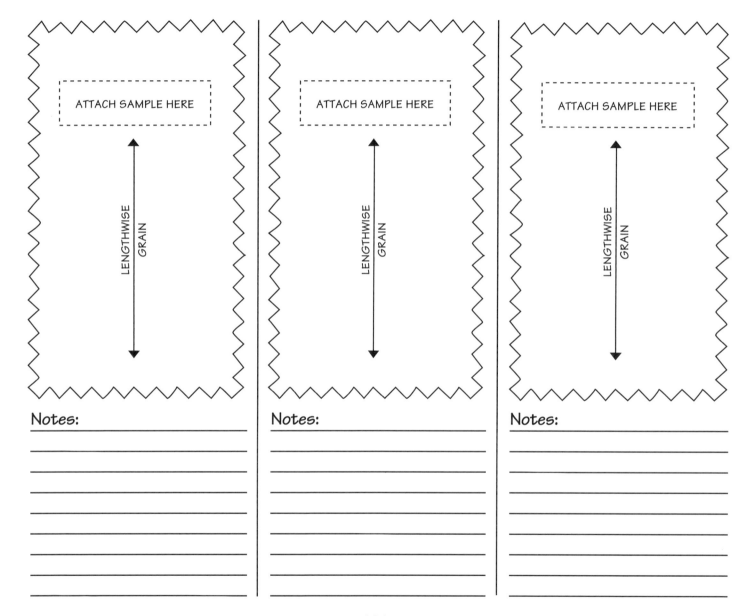

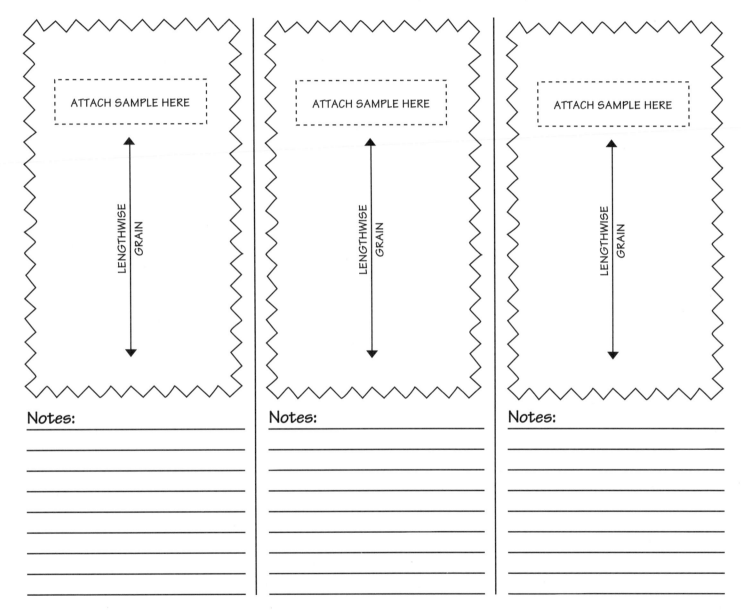

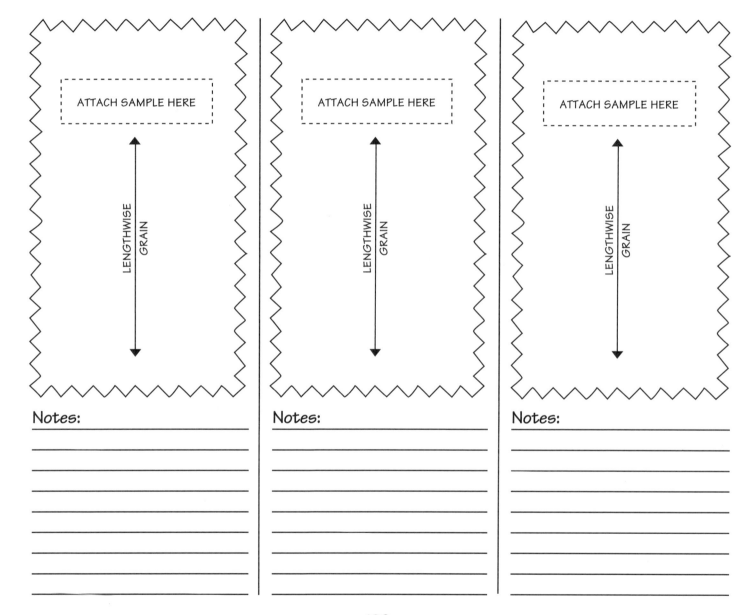

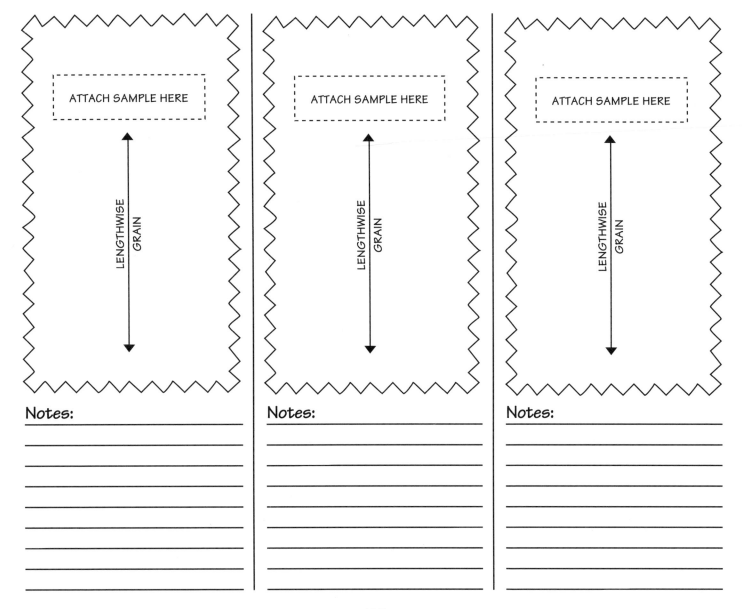

MAIL-ORDER SOURCES

If you can't find what you need at the local fabric store, you may want to explore mail-order sources. This is a good way to locate unusual or hard-to-find cottons.

Shopping by mail has limitations – it can be difficult to accurately judge a fabric's weight, hand, drape, pattern and overall appearance by inspecting a sample, some of which are ridiculously small. Many fabrics sell out quickly and usually can't be reordered, a problem if you're trying to coordinate several fabrics. Colors and fabric numbers sometimes get mixed up, especially when the source sends a lump of loose samples and skimps on the written descriptions. Some companies go out of business in a matter of months.

The best companies understand these limitations and work hard to explain their products and organize their catalogs. Some of them will even match threads, buttons and linings at your request.

There are three types of sources: fabric clubs, mail-order companies and swatching services. Most of them advertise in the major sewing magazines.

Fabric clubs: Fabric clubs charge an annual fee ranging from $10 to $50 or more. Catalogs and samples arrive in the mail at regular intervals.

Mail-order companies: Customer pays a deposit to receive a set of samples in the mail. The deposit is usually refunded when the samples are returned or if the customer places an order. Some companies specialize in cotton; others carry all kinds of fabrics.

Swatching services: Customer receives a small selection of fabric samples, based on a description of the fabric she seeks. This service is sometimes free and sometimes not. Some fabric clubs provide swatching services free to their members.

Getting good service

Most of your correspondence will be by mail or phone. Bad handwriting and muddled phone messages are common complaints of mail-order companies. It helps if you print your messages in easy-to-read block letters and provide your full name, address and phone number, in case there are any questions.

When leaving a phone message, speak slowly and clearly, spell your name and include the area code with your phone number, even if you live in the same area. State the reason for your call and suggest a good time for them to call you back. Otherwise, you'll wind up playing an endless game of telephone tag.

Sources of cotton

Baltazor's Fabrics and Laces
3262 Severn Avenue, Metairie, LA 70002
Delicate laces and fine cotton fabrics, including batiste, broadcloth, lawn, voile, organdy and piqué, plus hard-to-find Sea Island cotton. Catalog available for a small fee. Phone: 504-889-0333. Web site: www.baltazor.com.

Banksville Designer Fabrics
115 New Canaan Ave., Norwalk, CT 06850
Extensive collection of domestic and imported fabrics of all fibers. Mail-order and swatching services available. For more information, send self-addressed stamped envelope. Phone: 203-846-1333.

Batiks Etc.
411 Pine Street, Fort Mill, SC 29715
Authentic hand-waxed batiks and hand-painted fabrics from around the world, plus an interesting selection of handwoven fabrics, such as ikats and

stripes. Small annual fee for seasonal mailings. Toll-free: 800-228-4573. E-mail: batiks@cetlink.net.

Cotton Clouds
5176 South 14th Avenue
Safford, AZ 85546-9252

Cotton fiber and yarns in wide range of colors for spinning, weaving, knitting and crocheting, plus kits, patterns, books, videos, tools and equipment for spinning and weaving. Wholesale discounts available to qualified buyers. Catalog and samples available. Toll-free: 800-322-7888.

The Cotton Club
P.O. Box 2263, Boise, ID 83701

All-cotton prints and solids for quilting and other projects. Choice of three fabric clubs. Phone: 208-345-5567. Fax: 208-345-1217. Web site: www.cottonclub.com. E-mail: cotton@micron.net.

The Cotton Patch, 1025 Brown Avenue
Lafayette, CA 94549

Hundreds of cotton prints and solids for quilting and other projects, plus books and other supplies. Fabric samples available for a small fee. Catalog is free. Toll-free: 800-835-4418. E-mail: CottonPa@aol.com.

Dharma Trading Co.
P.O. Box 150916, San Rafael, CA 94915
1209 Third Avenue, San Rafael, CA 94901

Ready-to-dye fabrics, clothing blanks and supplies for textile artists. Has bleached and unbleached versions of some fabrics. Cotton inventory includes fleece, French terry, duck, gauze, interlock, jersey, poplin, ribbing, sheeting and others. Free catalog is informative and fun to read. Toll-free: 800-542-5227. Phone: 415-456-7657. Web site: www.dharmatrading.com.

Exquisite Fabrics, 1775 K Street NW, Washington, D.C. 20006

Fabrics imported from France, Switzerland and Italy, including couture cottons, silks, wools, bridal fabrics and laces. Custom sampling is free to customers who send specific details about the type of fabric they seek, including color, weight, garment style and yardage requirements. Phone: 202-775-1818.

Fabric Complements, 7689 Lakeville Highway, Petaluma, CA 94952

Extensive collection of high quality fabrics of all fibers, including cotton knits, prints, shirtings, piqués and others. Annual fee for seasonal mailings is higher than most other fabric clubs, but you get what you pay for. Samples are oversized and come mounted on 8 1/2- X 5 1/2-inch cards with suggestions for coordinating fabrics, garments, styles and methods of care. Insight newsletter keeps you up-to-date on fashion trends, and a complete inventory list is included with each mailing. The service is friendly and fast. Toll-free: 800-828-6561. Fax: 707-778-1729.

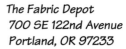

The Fabric Depot
700 SE 122nd Avenue
Portland, OR 97233

Extensive selection of popular fabrics of all fibers. Wholesale discounts of 40 percent to 50 percent available on full bolts. Also handles mail orders and special requests. Large warehouse-style store is worth a visit. Toll-free: 800-392-3376. Fax: 503-252-9556. Web site: www.fabricdepot.com.

G Street Fabrics Mail Order Service
12240 Wilkins Ave., Rockville, MD 20852

Extensive collection of better quality fabrics of all fibers. Annual fee is higher than most other fabric clubs, but you get 60 samples in each of 12 portfolio mailings. G Street also offers two sample sets that are mostly cotton – the Quilters Collection and the Children's Collection – each for a small fee. Custom sampling is free, but be specific about type of fabric, fiber, color, weight, garment style and yardage needed. Also sells patterns, notions and books. Two-floor store in Rockville is worth a visit. Toll-free: 800-333-9191. Phone: 301-231-8960. Fax: 301-231-9155.

Homespun, P.O. Box 4315
Thousand Oaks, CA 91359

Ten-foot-wide cotton fabrics for seamless draperies, table cloths, bedspreads and upholstery. Catalog and samples available for a small fee. Toll-free: 888-543-2998.

Keepsake Quilting
Route 25B, P.O. Box 1618
Centre Harbor, NH 03226-1618

Extensive selection of quilting fabrics and

supplies, including templates, kits, books, patterns and cutting tools. Bills itself as the largest quilting store in America, with more than 6,000 bolts of fabric. Monthly mailings are available in several different groupings for small annual fees. Write for more information and a free catalog, or visit their store, where you can admire (and buy) hundreds of handmade quilts from all over the United States.

Mill End Store, 9701 SE McLoughlin Blvd. Portland, OR 97222
Extensive selection of fashion fabrics of all fibers, plus home-decorating fabrics and upholstery. Plain and fancy cottons are abundant. For information on mail-order services, send a long, self-addressed, stamped envelope. Large warehouse-style store is worth a visit — tour buses are common. Discounts available to qualified buyers. Phone: 503-786-1234. Fax: 503-786-2022.

Philips Boyne Corp., 135 Rome Street Farmingdale, NY 11735
Large collection of imported and domestic cotton shirtings, including broadcloth, oxford cloth, pinpoint oxford, novelty weaves, tone-on-tone and yarn-dyed stripes, checks and plaids. Small fee for set of samples. Toll-free: 800-292-2830.

Rupert, Gibbon & Spider, Inc. P.O. Box 425, Healdsburg, CA 95448
Organic cottons, undyed fabrics, dyes, paints and supplies for textile artists. Free catalog. Samples available for a small fee. Discounts on full bolts. Toll-free: 800-

---

> **Not an endorsement**
> The purpose of this list is to provide sources of cotton fabrics. Although we would not knowingly include a firm with disreputable business practices, inclusion on this list is not intended to be a recommendation or endorsement of any kind.

442-0455. Fax: 707-433-4906.

Sawyer Brook Distinctive Fabrics P.O. Box 1800, Clinton, MA 01505-0813
Small annual fee for seasonal mailings of fine fabrics imported from Italy, France and elsewhere. Specializes in fabrics made from natural fibers, especially fine cotton prints. Carries Liberty of London and Viyella cottons. Oversized samples are nicely presented and packaged with care so you won't get them mixed up. Phone: 978-368-3133. Fax: 978-365-1775. Web site: www.sawyerbrook.com.

Sew Natural, 45551 Stoney Run Drive Great Mills, MD 20634
Specializes in natural fibers, including all-cotton interlock, jersey, French terry, denim, twill, sheeting and flannel. Catalog and seasonal mailings available for small annual fee. Write for more information.

Testfabrics, Inc. 415 Delaware Avenue West Pittston, PA 18643
Specializes in undyed, unprinted fabrics of all fibers for textile artists. Carries a wide variety of different types of cotton fabrics, all of which are white or natural in color. Catalog available. Call or write for more information. Phone: 717-603-0432.

Vreseis Ltd., P.O. Box 87 Wickenburg, AZ 85358-0087
Cotton fibers, yarns and fabrics made from FoxFibre®, naturally colored cottons developed by Sally Fox. Colors are earthy browns and greens, along with whites and ivories. Most FoxFibre® is organically grown and sold to clothing manufacturers, but some yardage is available. Samples can be had for a small fee. Write for more information, but be prepared to wait for a reply — the demand exceeds the supply.

Zoodads, P.O. Box 15073 Riverside, RI 02915
Mostly cotton children's fabrics and juvenile prints, including a nice selection of fabrics for boys. Small annual fee for seasonal mailings. Phone: 401-437-2470. Web site: www.zoodads.com.

Another resource

The Fashiondex, Inc., 153 West 27th St., New York, NY 10001
The **Fashiondex** is a resource guide for the apparel industry that is updated at regular intervals. It contains wholesale sources of fabrics, buttons, notions, trims, belt buckles and other findings, plus lots of information useful to small manufacturers, independent designers and entrepreneurs. Phone: 212-647-0051. Fax: 212-691-5873. For a peek at what's in the Fashiondex, check out their web site at www.fashiondex.com.

LOSSARY OF COTTON TERMS

African cotton: Cotton grown in Africa. The quality tends to be inferior to U.S. cotton.

algodón: Spanish for cotton.

American cotton: Any cotton grown in the United States, although it usually refers to Upland cotton, which makes up the bulk of the U.S. crop.

American-Egyptian cotton: Another name for Pima cotton, which was developed from Egyptian seed.

American Peeler cotton: Another name for Peeler cotton.

Andes cotton: Another name for Peruvian cotton.

Anguilla cotton: Very fine cotton grown on the island of Anguilla in the West Indies. It is believed that seeds from Anguilla were planted in the Sea Islands in the 1870s, making it the source of Sea Island cotton.

Arizona cotton: Another name for Pima. Also called Arizona-Egyptian cotton.

Ashmouni: Variety of Egyptian cotton grown in the Upper Nile Valley.

basting cotton: Cotton sewing thread used for basting or holding material in place before the final stitching. It is not as strong as regular sewing thread.

Baumwolle: German for cotton.

black leaf: Weather-damaged cotton that contains pieces of black leaves. Usually describes a variety of Indian cotton, which is called red leaf when it is not damaged.

black seed: Wide variety of American cottons that produce smooth black seeds.

Blue Bender cotton: Inferior cotton with a bluish color that cannot be bleached out using normal processes. It grows wild along the Mississippi River and its tributaries.

blue cotton: Unusually white cotton.

blush: Desirable creamy white color and luster of certain cottons.

bombax cotton: Vegetable down or floss produced by various tropical trees and plants. The fiber is not very strong and is used mainly for stuffing and wadding, although it is sometimes mixed with other stronger fibers and spun into yarn.

book cloth: A variety of plain-weave, coarse cotton fabrics, including osnaburg, flat duck and sheeting, that are filled with starch and clay or otherwise coated with a special finish and sometimes embossed to simulate leather. Used to cover and bind books. Also called binding cloth.

Brazilian cotton: Any cotton from Brazil, most of which is good grades of American Upland varieties.

bread-and-butter cotton: Medium-quality cotton fibers that are always in demand.

brushed twill: A very soft, comfortable cotton twill with a brushed surface similar to flannel, but less pronounced.

buckram: A coarse, open fabric made with a loose plain weave, that is heavily sized. Used as a stiffener to give shape to a garment.

bumblebee cotton: Has very short fibers.

bump cloth: Coarse cotton cloth made with bump yarns, which are made from cotton waste.

buttery cotton: Cotton with a creamy, beige or light brown color.

buzz-fuzz cotton: Cotton with extremely short fibers.

cellulose: A carbohydrate found in organic woody substances of most plants. It is the basic substance of all vegetable fibers – cotton, linen, ramie, jute and hemp.

challis: Very soft fabric made with a plain weave or sometimes a twill weave and usually printed with small floral patterns or paisley designs. The fine, lightweight fabric has a smooth surface that may be lightly napped to increase its soft hand. It is sometimes mercerized to add strength and luster. Challis is one of the softest cottons, but it is not very common. Rayon and wool versions are easier to find.

chemical cotton: Purified and bleached cotton linters that are used as a source of cellulose for some types of rayon, cellulose acetate and nitrocellulose lacquers.

Chinese cotton: The species *Gossypium herbaceum* forms the major part of the Chinese cotton crop. The coarse wiry fiber is whiter and cleaner than Japanese or Indian cotton, but it averages only about ½ inch in length. Most of it is consumed locally or mixed with other fibers.

coton: French for cotton.

cotone: Italian for cotton.

cotton cashmere: Misleading term used by the fashion industry to describe very soft cotton fabrics that contain no cashmere.

cotton grading: A system of grading cotton according to its growth, maturity, cleanliness, color and freedom from insect damage. It does not refer to the staple length. The classifications include: 1. Good middling; 2. Strict middling; 3. Middling; 4. Strict low middling; 5. Low middling; 6. Strict good ordinary; 7. Good ordinary. There are also five half-grades. As the term indicates, middling is the middle or basic grade and is the grade upon which market quotations are based.

cotton grass: Fiber from the fruit of the reed mace, used to stuff upholstery.

Cotton Incorporated: U.S. organization that promotes the use of cotton fabrics and other cotton products.

cotton picker: Machine used for picking cotton.

cotton-rich: Fabric containing at least 60 percent cotton fibers. The other fiber is usually synthetic. These fabrics often carry a Natural Blend® label, a registered trademark of Cotton Incorporated.

cotton tree: Wide variety of trees from the genus *Bombax*, *Ceiba* and *Eriodendron*, which produce soft, fluffy cotton-like fibers.

cotton waste: Waste that accumulates during the manufacturing process. Soft cotton waste is used for padding. Hard waste is used to make paper and other items. Spinnable waste is mixed with new cotton or other fibers for use in weaving or knitting. Non-spinnable waste is used to make mops, cords and other items.

cotton wax: Natural wax-like substance present in small quantities on the raw cotton fiber, making it water resistant. It is removed by scouring.

cotton wool: Sometimes used to describe raw cotton.

cottonwood: Soft fibrous bark from the cottonwood tree of the genus *Populus*, in the United States. It has been used by Indian tribes to make ropes, garments and other items. It is not the same as the cotton tree or silk cotton tree.

cut staple: Inferior cotton fiber that has been accidentally cut during ginning because the fiber was too damp. The term is sometimes used incorrectly to describe man-made staple, especially rayon staple.

Dacca muslin: Famous, handwoven cotton muslin made in Dacca, India (now part of Bangladesh). It is considered to be the finest of the Indian muslins and has been made for centuries by the people of Dacca, from locally grown cotton. Ten yards of the fabric may weigh as little as 3 or 4 ounces.

delinting: The process of separating the very short fibers, called linters, from the seeds of cotton. Also called "second-time ginning" of cotton. The linters are used to make rayon and acetate.

dobby weave: Figure weave produced by an attachment on the loom that weaves an extra set of yarns into a simpler background weave. Fabrics usually have small bird's eye, dot or other patterns.

dogs: Very poor quality of raw cotton.

double carded: Cotton carded twice to improve its quality. Yarns spun from such cotton are superior to single carded yarns, but inferior to combed yarns.

double combing: Cotton combed twice to produce very high quality, very fine yarns.

doubles: Used to describe two-ply cotton yarns, especially in the United States.

durable crease: Pressed pleats or creases that stay sharp when a garment is laundered. Ironing is not required. Not the same thing as durable press.

end: Lengthwise yarns on a loom or in a fabric.

extra white cotton: Color classification of cotton that is brighter and less yellow than the white group.

filling: Crosswise yarns on a loom or in a fabric.

flame resistant: Fabric that burns but self extinguishes rapidly when the flame is removed. It is not the same as fireproof, which means the fabric will not burn.

flax: Name of the plant that yields a bast fiber from its stalks, better known as linen.

Flax is the oldest textile fiber known and is currently produced in Belgium, Ireland, the Netherlands, France, Italy, Egypt and the countries that formerly belonged to the Soviet Union. The fiber is very strong and has a pronounced luster, but it wrinkles easily and lacks elasticity.

French cotton: Lustrous floss from the fruit of the plant *Calotropis gigantea*. The stem produces a hemp-like fiber.

gin: Machine used to separate cotton fiber from seeds. There are two types: the saw gin and roller gin.

ginning: The process of separating cotton fiber from seed. The three main products of ginning are lint, which is spun into cotton yarns; linters, which are used to make rayon, acetate and other products; and cottonseed, which is used to make cottonseed oil and fertilizer.

Giza: Variety of Egyptian cotton.

Gossypium: The genus of plants that produce cotton fibers. There are several species, including *Gossypium arboreum*, *Gossypium barbadense*, *Gossypium herbaceum*, *Gossypium hirsutum* and *Gossypium religiosum*.

gray cotton: Stained gray by exposure.

green cotton: Immature cotton that is unusually damp because it was picked before the boll was ripe.

harness: A frame that holds the heddles in position on the loom during weaving. The heddles are a series of wires that keep the warp yarns under control at all times. Looms may have more than one harness, which is also called a shaft.

hoa mein: Chinese for cotton.

hull fiber: Sometimes used to refer to the shortest linters.

immature cotton: Undeveloped, immature cotton fiber. The weak brittle staple does not spin or dye satisfactorily. Also called dead cotton and unripe cotton.

Jumel cotton: Perennial tree cotton named for the French engineer who noticed it in a garden in Egypt and suggested it for commercial cultivation. This marked the beginning of Egypt's reign as a producer of fine cottons. Jumel cotton was cultivated from about 1820 to 1860, when it was replaced by improved types.

King cotton: Famous variety of Upland cotton that originated in North Carolina in about 1890. It has been replaced by King Improved, King Gold Dust and other improved varieties.

leafy cotton: Ginned cotton that contains a large amount of small leafy matter.

lint: Raw cotton that has been ginned. Also used to describe loose short fibers and soft fluff produced by fabrics or garments when they are laundered.

linters: Short cotton fibers that remain stuck to the seeds after the cotton is ginned. Linters usually are less than 1/8 inch in length. They are removed from the seeds and used to make rayon fabrics and to stuff upholstery and mattresses.

Maco cotton: Tree cotton with smooth, strong, lustrous long fiber, cultivated mainly in Brazil. It was developed from a cross between an Egyptian tree cotton and Sea Island cotton. The trees bear cotton for seven to 15 years or more, and can produce two crops a year in ideal conditions. Also spelled Moco.

meaty cotton: Clean cotton that produces little waste when it is spun.

mercerized sewing thread: Boilfast, three-cord cotton thread that has been treated with caustic soda (mercerized) to give it strength, luster and an affinity for dyes. Available in a wide range of colors.

middling: The basic cotton grade. Middling cotton is creamy white and contains only a few pieces of leaf and immature seeds.

Mit Afifi: Fine, strong Egyptian cotton with a lustrous dark brown, extra-long fiber, named for the village where it was discovered in the 1880s. Sakellarides and American-Egyptian are two hybrids derived from Mit Afifi.

mock-Egyptian cotton: Cotton yarn or fabric that is tinted a light buff shade to imitate Egyptian cotton.

modified cellulose fiber: Cotton fibers treated with caustic soda to change their chemical and physical properties in order to increase strength, luster and affinity for dye.

momen: Japanese for cotton.

National Cotton Council: U.S. organization that promotes the use of cotton fabrics and other cotton products.

Natural Blend®: Registered trademark of Cotton Incorporated for easy-care fabrics that contain at least 60 percent cotton.

P/C, P-C: Abbreviation for blend of cotton and polyester.

Peeler cotton: Upland cotton with a staple length of 1¹/₈ inches or more.

pick: Another name for crosswise yarns.

piece: Standard length of continuous cloth taken from the loom, generally about 80 to 100 yards. The pieces may be sewed together end-to-end.

piece dyed: Fabrics that are dyed, usually a solid color, after they have been woven or knitted. Iridescent effects, stripes and other patterns may be produced on fabrics made of blends of fibers that respond differently to the same dye.

piece goods: Fabrics woven in shorter lengths to be sold by the yard in retail stores. The term also is used to describe all fabrics that have not been cut. Also called yard goods.

pillow tubing: Cotton fabric woven in the form of a tube or stitched together at the selvages to form a tube. It is used to make pillowcases.

pillowcase linen: Smooth bleached linen or cotton fabric made with a high count, tight plain weave.

progressive shrinkage: Irreversible type of shrinkage that is unique to wool and other fibers that have scales. When the fiber is subjected to heat, steam, pressure and friction, the scales become interlocked and cannot be untangled.

ply yarn: Yarn made by twisting together two or more single strands. A two-ply yarn has two individual strands, a three-ply yarn has three strands and so forth. The individual strands can be counted when the yarn is untwisted. When a single yarn is untwisted, the fibers separate and come apart. Ply yarns are smoother, stronger, heavier and more uniform than singles.

raw cotton: Cotton fiber in its natural state, before it has been ginned.

reginned cotton: Cotton that is ginned a second time to remove defective fibers.

residual shrinkage: Another name for relaxation shrinkage, which occurs when the yarns relax the first time a fabric is washed.

reverse blend: Blend that contains a greater percentage of natural fiber than synthetic, usually of cotton or wool and polyester. The yarn is spun so that the natural fiber wraps around the synthetic fiber, giving the fabric an all-natural look and feel. The hidden synthetic fiber adds strength and durability.

ripe cotton: Fully mature cotton that is ready to be spun into yarn.

run-of-the-mill: Any fabric or yarn that is not inspected or does not need inspection. The product is usually classed as seconds because it does not meet standards for quality and may contain defects.

run-of-the-loom: Fabric that is ready for shipment as it comes from the loom. The fabric is not inspected and weaving defects are not corrected.

Sakellarides: Cream-colored Egyptian cotton with very fine, extra-long staple and uniform length and thickness. Named for a Greek merchant who discovered the plant. It was the outstanding variety of Egyptian cotton for many years, but it has been replaced by improved strains, including Maarad and Giza.

seed cotton: Any cotton that contains seeds, after it has been picked but before the seeds have been removed in the gin.

sheeting: Plain-weave fabric, usually made of carded yarns and used to make sheets.

silk cotton: General term for a variety of fine lustrous fibers obtained from the seed pods of many plants and trees, especially the *Bombaceae* family. The short fibers are used mainly for stuffing because they are difficult to spin. Kapok and bombax are the most common commercial varieties.

silverfish: Silvery, wingless, cotton-eating insects with long feelers and bristles on the tail, usually found in dark and damp places. They especially like starched and sized cotton.

singles: Used widely in the United States for a single yarn, especially when it is cotton. A ply yarn is made from singles.

soft cotton: Fibers that are smooth, fairly clean and soft, such as Pima, Egyptian and Sea Island cottons.

S x P cotton: American-Egyptian cotton developed from a cross between Pima and

Egyptian Sakellarides. It is a little coarser than Pima, lighter in color and has shorter staple (about 1 1/2 inches), but produces a higher yield per acre and stronger yarn.

staple fiber: Manufactured filament fibers that have been deliberately cut into shorter lengths to be spun into yarn.

straw cotton: Heavily sized cotton thread that feels like straw, used to make straw-like products.

tow: Short tangled bast fibers, such as flax (linen), that are less than 10 inches long. Used to make low-grade yarn and twine.

tree cotton: Cotton from *Gossypium arboreum* and certain South American perennials that grow to be large shrubs or small trees.

true percale: Smooth, luxurious percale sheeting fabric made with combed cotton and about 200 yarns per square inch.

utility percale: Percale sheeting fabric made with carded cotton and about 180 yarns per square inch. The yarn is finer than the yarn used in muslin sheeting, but not as fine as the yarn in true percale. The fabric is used by the garment industry to make all types of clothing and sheets.

Uppers: Cotton grown in the Upper Valley of the Nile River in Egypt. Uppers cotton is usually of the Ashmouni variety, and has shorter staple than cotton from the Delta region of Egypt. It is not the same as Upland cotton, an American variety.

warp: Lengthwise yarns on a loom or in a fabric, so-called because they are controlled by the warp beam, a part of the loom.

weft: Technically accurate name for the crosswise yarns on a loom or in a fabric. Filling is the preferred term because weft is too easily confused with warp.

yarns: The textile industry's term for the individual threads on a loom or in a fabric, including, but not limited to, thick yarns used to knit sweaters.

LOSSARY OF FINISHES

beetling: A process used on cotton and linen cloth to flatten it and increase the luster. Damp fabric is hammered with heavy wooden mallets as it passes over a cylinder.

bluing: The process of "bleaching" a yarn or fabric by dyeing it with a reddish-blue or blue color to neutralize its yellowish tint.

boiling off: Cotton fiber or fabric is boiled in a solution of caustic soda, soda ash, soap or synthetic detergent to remove the natural waxes and gums, which improves cotton's ability to absorb moisture. Also called scouring.

calendering: A finishing process applied to fabrics by passing the cloth between one or more rollers, called calenders, usually with carefully controlled heat and pressure, to produce a variety of surface effects. Moiré and glazed finishes are examples.

cambric finish: A finish that combines singeing and calendering to produce a firm bright finish.

closing: Loosely woven cotton fabrics that have been sized to fill in the open spaces between the yarns. After sizing, the fabric is calendered to flatten the yarn, forming a compact cloth.

coated fabric: Any fabric that has been coated with a substance to make it more durable or water repellent, among other characteristics.

converted fabric: A term used to describe any fabric that has been finished in some way, to distinguish it from gray goods.

embossing: A finishing process used to add texture or raised designs to fabric, produced by passing the fabric through a series of engraved rollers, using heat and pressure.

filled cloth: A cotton fabric that has been coated with starch, clay or another substance to close the spaces between the threads. The filling increase the fabric's weight and alters its appearance, making it look more closely woven.

flame retardant: A chemical applied to fabric to make it flame resistance.

glazing: A process that produces a smooth high polish on the fabric surface. The cloth is coated with starch, paraffin, shellac or synthetic resin and run through a friction calender.

holland finish: Applied to sheer cotton fabrics to make them more opaque. Named for an obsolete linen fabric called holland.

imitation linen finish: Used to describe cotton fabric that has been beetled or otherwise finished to look like linen.

lacquer finish: A chemical treatment that forms a thin, smooth, highly glazed film on the surface of the fabric. It is sometimes applied in patterns, producing a dull-and-lustrous design on the fabric.

lacquer prints: Fabrics printed with lustrous colored lacquers.

lustering: Any finishing process that produces a luster on yarn or fabric by heat, pressure, steam and friction calendering.

moiré: A popular finish that resembles a watermark or stain on the fabric. It is applied to cotton, rayon, acetate, silk and other fibers by pressing the fabric between engraved rollers. The pressed area reflects light differently than the unpressed area.

nainsook finish: A soft finish that produces a slight luster on both sides of the fabric.

napped fabric: Fabric that has been finished to raise the nap on one or both sides. The nap may cover the entire fabric surface or form subtle stripes or figures.

organdy finish: A crisp-to-stiff finish given to fine sheer fabrics of cotton, silk or synthetic fibers. It may not be permanent.

parchmentizing: A type of finish used on cotton fabrics to produce a variety of effects, including transparency, increased texture or a linen-like hand.

permanent starchless finish: Crisp finish that holds up to repeated launderings and can be restored by ironing the fabric.

prewashing: A finishing wash used to produce a softer hand and a worn appearance, especially on denim and corduroy.

pure starched finish: A finish applied to cotton fabrics to produce a crisp hand.

sateen finish: A highly lustrous finish with a fairly crisp hand, used on cotton fabrics to imitate satin. When used on other fibers, it is called a satin finish.

schreinering: A finish used to produce a high luster on cotton fabric. The fabric is flattened between a smooth roller and a roller that is engraved with fine parallel lines. The process simultaneously flattens the threads and produces a series of ridges that reflect light, but are not visible to the naked eye.

singeing: A finishing process that removes lint, fuzz and other protruding matter from a fabric's surface by passing the fabric quickly over a jet flame.

slack mercerization: A process used to produce a crosswise stretch on cotton and cotton/blend fabrics. The fabric is immersed in a sodium hydroxide bath to shrink it. It is called slack mercerization because no tension is applied to the fabric during the chemical bath.

soft finish: Any finish that produces a soft pleasant hand on fabric.

souring: A finishing treatment that uses a weak acid solution to neutralize residual alkali in textile materials, especially cotton.

starching: A finish applied to cotton fabric that combines starch and calendering to add body and stiffness and to improve the fabric's appearance.

suiting finish: A finish given to cotton fabrics to imitate linens.

Swissing: A type of calendering finish that compresses cotton fabric to produce a smooth compact appearance with moderate luster.

taffetized fabric: A cotton fabric finished with a partial glaze to imitate taffeta.

THPC cotton: Cotton that has been chemically treated to make it flame- and char-resistant.

Tris: Common name for a compound known as trist (2-, 3-, dibromopropyl) phosphate. It was used to make children's sleepwear flame-retardant, but it was linked to cancer and has been banned since 1977.

weighting: The process of adding weight or body to a fabric or yarn by the addition of various materials, such as starch, clay or sizing. It usually refers to silk, but cotton, wool and rayon fabrics are also weighted. Weighting is usually used to disguise the quality of inferior fabrics.

zinc finish: A type of calender finish that uses a zinc rather than a copper roller. The effect is similar to regular calendering.

BIBLIOGRAPHY

"A Dictionary of Textile Terms." 13th ed. New York: Dan River Inc., 1980.

Anderson, Enid. "The Spinner's Encyclopedia." Great Britain, David & Charles, 1987.

Corbman, Bernard P. "Textiles: Fiber to Fabric." 5th ed. New York: McGraw-Hill, Inc., 1975.

"Cotton Fabrics Glossary." Boston, New York & others: Frank P. Bennett & Co., Inc., 1914.

"Encyclopedia of Textiles." Englewood Cliffs, New Jersey: Prentice-Hall, Inc., 1960.

Flusser, Alan. "Clothes and the Man: The Principles of Fine Men's Dress." New York: Villard Books, 1989.

Gioello, Debbie Ann. "Profiling Fabrics: Properties, Performance and Construction Techniques." New York: Fairchild Publications, 1981.

_____. "Understanding Fabrics: From Fiber to Finished Cloth." New York: Fairchild Publications, 1982.

Hardingham, Martin. "The Fabric Catalog." New York: Simon & Schuster Pocket Books, 1978.

Hollen, Norma and Jane Saddler and others. "Textiles." 6th ed. New York: Macmillan, 1988.

Houck, Catherine. "The Fashion Encyclopedia." New York: St. Martin's Press, 1982.

Joseph, Marjory L. "Essentials of Textiles." 2nd ed. New York: Holt, Rinehart and Wilson, 1980.

Lyle, Dorothy Siegert. "Modern Textiles." New York: John Wiley & Sons, Inc., 1976.

McRae, Bobbi A. "The Fabric and Fiber Sourcebook: Your One-and-Only Mail-Order Guide." Newtown, Connecticut: The Taunton Press, Inc., 1989.

Miller, Edward. "Textiles: Properties and Behaviour in Clothing Use." London: B.T. Batsford, 1989.

Pizzuto, Joseph J. "Fabric Science." 5th ed. revised by Arthur Price and Allen C. Cohen. New York: Fairchild Publications, 1987.

Ross, Mabel. "The Encyclopedia of Hand Spinning." Loveland, Colorado: Interweave Press, Inc., 1988.

Shaeffer, Claire. "Fabric Sewing Guide." Radnor, Pennsylvania: Chilton Book Company, 1989.

"Textile Handbook." Washington, D.C.: American Home Economics Association, 1964.

Wingate, Isabel B. "Fairchild's Dictionary of Textiles." 6th ed. New York: Fairchild Publications, 1984.

_____. "Textile Fabrics and Their Selection." 7th ed. Englewood Cliffs, New Jersey: Prentice-Hall, Inc., 1976.

MAGAZINES AND PERIODICALS

Sew News, PJS Publications, Peoria, Illinois.

Sewer's SourceLetter, CraftSource, Seattle, Washington.

Threads, The Taunton Press, Newtown, Connecticut.

PROMOTIONAL SOURCES

Cotton Incorporated, New York, New York. Promotional brochures and 1991 annual report.

Lands' End mail-order catalog, Volume 28, April 1992.

National Cotton Council of America, Memphis, Tennessee. Promotional material, samples of cotton and cotton fabrics.

INDEX

A
absorbency 6, 10, 94
A.C.A. ticking 96
acetate 8, 19
acid-washed denim . 46
acrylic 8, 19
animal fibers 8
army duck 54
army oxford 80
awning stripe 32

B
basket weave 73
bast fibers 8
batik 24-25
batiste 26-27
Bedford cord 82
bird's eye piqué 82
biscuit duck 54
bishop's lawn 68
bleach 6, 7, 12, 16, 20, 21
blends 65
blister crêpe 84
boatsail drill 32
Bohemian ticking 96
boll weevil 14
bouclé yarn 36
British poplin 86
broadcloth 28-29
brocade 44
broken twill 99
brushed cotton 58
brushed denim 46

C
calico 30-31
calendaring 17
cambric 25
candlewick fabric 36
candlewick yarn 36
canvas 32-33
carded cotton 10, 15
caustic soda 85
caustic soda crêpe . 85
cellulose 8, 9
cellulosic fibers 8
chambray 34-35
chambray gingham .. 62
cheesecloth 60
chenille 36-37
chenille yarns 37
chino 38-39
chintz 40-41
clipped spots 48
colored cotton 79
combed cotton 10, 15, 26
converter 78
corduroy 42-43
corduroy toweling 95
corkscrew yarn 36

cotton
 boll 9, 10, 14
 fiber 6, 9
 gin 13
 plant/bush 9
 seed 9
cotton/polyester
 blends 65
crêpe 84
cretonne 40
crinkle cloth 90

D
damask 44-45
damask ticking 96
denim 46-47
desizing 16
diaper flannel 58
dimity 102
dotted Swiss 48-49
double cloth 50
double damask 44
double jersey 50
double knit 50-51
double-faced knit 50
double-wide cloth 74
drapeability 7, 10
drill 52-53
dry cleaning 20, 101
dry cleaning solvent . 12
duck 54-55
durability 10, 11
durable press 92
dyeing 16, 24

E
Egyptian cotton 12, 14, 28, 29, 69, 92
elasticity 11
embossed fabric 84
embroidery 56
end-and-end 34
even twill 52
eyelet 56-57
England 12

F
fashion trends 17
fiber structure 9
finish, types of
 flame-retardant .. 59
 crisp 77
 durable press 92
 glazed 40
 permanent 17
 semi-durable 17
 shrink-resistant .. 39
 stain-repellent 41
 temporary 17
flammability 12
Flammable Fabrics Act
 59
flannel 58-59
flannelette 58
flat duck 54
flexibility 10
flocked fabrics 48
Fox, Sally 79

G
gabardine 98
gauze 60-61
genetically engineered cotton 78
Genoa cord 42
gingham 62-63
ginning of cotton 15
glass toweling 94
grading of cotton 15
grain of fabric 18
greige goods 78

H
harvesting cotton 15
 by hand 15, 29
 by machine 15
herringbone 98, 99
hopsacking 72
huck toweling 94

I
ikat 24
Indian cotton .. 12, 14, 71
indigo dye 46
Industrial Revolution 13
interlock knit 64-65

J
jacquard weave 45
jean 46
jersey 66-67

K
kapok stuffing 96
khaki 38
knit fabrics 51
knit terry 94
knitting 16

L
lace 56
lappet weave 57
lawn 68-69

laundering cotton 6
laundry detergent ... 20
leaf fibers 8
leno weave 60, 61
leotard fabric 64
Levi Strauss 17, 47
Liberty of London 68
linen 6, 8, 19
lint 36
linters 17
longcloth 68
looped pile 95
luster 88
lyocell 8

M

Madras 70-71
madras gauze 70
madras gingham 70
manufactured fiber ... 8
marlin 32
marquisette 60
matte jersey 66
mercerized cotton ... 25
Mexico 12, 14, 31
mildew 7, 12, 32
   prevention of 33
modacrylic 8
monk's cloth 72-73
muslin 74-75

N

nainsook 25
natural fibers 8
needlepoint 56
needlework 56
Nelo batiste 25

numbered duck 54
nurses' gingham 62
nylon 8, 19

O

olefin 8
organdy 76-77
organdy finish 76
organic cotton 78
organza 76
origin of cotton 12
oxford cloth 80-81

P

P.F.P. fabrics 78
parchmentizing 77
percale 74
perspiration 11, 20
Peruvian cotton ... 12, 14, 29
piece-dyed fabric 63
pigskin piqué 82
pile fabric 42, 94, 95, 100
Pima cotton 14, 28
pinpoint oxford 80
piqué 82-83
piqué voile 102
plain weave 35
plant fibers 8
plissé 84-85
ply yarns 32
polished cotton 40
polyester 8, 19
poplin 86-87
prewashed cotton 11, 21
prewashed denim 46

printing 16, 24, 30
protein fibers 8
puckered cloth 90

R

rayon 8, 19
rib knit 50
rib weave 87

S

sack cloth 72
sailcloth 32
sateen 88-89
satin weaves 89
scrim 60
Sea Island cotton ... 14, 28, 29, 31, 92, 93
seed fibers 8
seed voile 102
sheeting 74
shirting 92-93
shrinkage 7, 11, 21, 39
shrinkage control 39
silk 8, 19
single knit 51
sizing 17, 74
slavery 13
spandex 66
spiral piqué 82
spiral yarn 36
splash voile 102
stains, removal of .. 22
staple length 15
starch 80
static cling 7, 10
stone-washed denim 46
storage of cotton ... 22

strength 6, 7
stretch interlock 64
stretch jersey 66
stripes 28
stuffer yarns 82, 83
stuffing 96
Supima cotton 28
sweater knit 64
Swiss batiste 25
Swiss cotton 49
Swiss finish 76
Swiss organdy 76
Switzerland 49
swivel dots 48
synthetic fibers ... 6, 7, 8, 10

T

T-shirt knit 66
table damask 44
Tencel® 8
tentered cotton 16
terry cloth 94-95
terry velvet 94
thread count 74
ticking 96-97
tie dyeing 24
tissue gingham 62
triacetate 8
tube knit 50, 67
Turkish toweling 94
twill fabrics 98-99

U

uneven twill 52
unstable weaves 72
Upland cotton 31

uniform cloth 38
U.S. cotton crop 12, 13, 14, 31

V

vegetable dyes 70
velour 100
velour toweling 94
velvet 100
velveteen 100-101
velvet, cotton 100
velveteen plush 100
Victoria lawn 68
voile 102-103
voile yarns 102

W

waffle piqué 82
wales 42-43
warp sateen 88
warp yarns 16, 115
warp-faced twill 52
weaves . 35, 44, 45, 53, 60, 72, 73, 81, 87, 98
weaving 16
weft yarns 115
whipcord 98
Whitney, Eli 13
wool 6, 7, 8, 19
woven fabrics 16
wrinkles 7, 21
   resistance to 90, 91

Y

yarn-dyed fabric 62
yarn sizes 103

Z

zephyr gingham 62

OTHER BOOKS IN THE SERIES

For more information about books in Julie Parker's **Fabric Reference Series**, write to Rain City Publishing, P.O. Box 15378, Seattle, WA 98115-0378 or call 206-527-8778 weekdays from 8 a.m. to 5 p.m. Pacific Time. If you send us a fax, we'll fax back an order form. Our fax number is 206-526-2871.

VOLUME I
All About Silk:
A Fabric Dictionary & Swatchbook

92 pages plus 32 silk samples. Revised edition published in 1992. Fourth printing 1997. ISBN 0-9637612-0-X. LCCN: 93-117999r93.

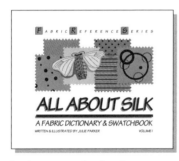

Brief introduction covers the history of silk, the main sources of silk and the silk textile industry, followed by descriptions and samples of 32 silk fabrics, in this order: batiste de soie, broadcloth, brocade, charmeuse, chiffon, China silk, cloqué, crêpe, crêpe de Chine, douppioni, faille, gabardine, georgette, habutai, jacquard, knit, matelassé, matka, noil, organza, peau de soie, pongee, printed silk, sandwashed silk, shantung, suiting, surah, taffeta, Thai silk, tussah silk, tweed, velvet.

All About Silk is extremely well organized, providing quick answers to hundreds of questions about an often confusing subject. Terms such as cultivated silk, pure silk, raw silk, reeled silk, spun silk and wild silk are cleared up once and for all.

The author uses simple drawings, an easy-to-read, consistent format and uncomplicated language to bring clarity to a subject that many of us don't understand. It should be required reading for anyone who loves silk!

VOLUME III
All About Wool:
A Fabric Dictionary & Swatchbook

144 pages plus 35 wool samples. Published in 1996. Second printing 1998. ISBN 0-9637612-2-6. LCCN 96-92209.

Comprehensive introduction covers characteristics of the fiber, history of wool, breeds of sheep, types of wool, the main sources of wool, the wool textile industry, new technology and wool quality, followed by descriptions and samples of 30 wool fabrics and five luxurious specialty hair fibers, in this order: blanket cloth, boiled wool, bouclé, cavalry twill, challis, coating, crêpe, Donegal tweed, double cloth, double knit, felt, flannel (woolen), flannel (worsted), gabardine, glen plaid, Harris tweed, herringbone, homespun, houndstooth, jacquard, jersey, loden cloth, melton, menswear suiting, novelty suiting, plaid, satin, tropical suiting, tweed, whipcord, alpaca, angora rabbit, camel's hair, cashmere, mohair.

All About Wool is packed with information about the yarns, weaves and finishing techniques used to make woolen and worsted fabrics. Terms such as virgin wool, lamb's wool, merino wool, superfine wool and reused wool are clearly explained. A special section describes similar fibers from other animals, such as camels and goats, making it the complete book of wool fibers and fabrics.